Super Max - Dipole Antennas

By Edward J. Shortridge W4JOQ

If you wish to have the absolute newest and the best "Inverted V or a Horizontal" antenna that outperforms others on the 75-80 or 160m bands, this book will introduce you to antennas that have the lowest and widest SWR bandwidths available. These antennas strive to be the most efficient, delivering your full transmitter power as RF gain. The large variety of designs will easily meet your needs, all of which are a part of a "Unified Broad-Banding" method.

Disclaimer

Most of the antennas presented in this book consist of totally new concepts in the art of antenna design and they use a relatively new Unified Broad-Banding method in order to tie them all together.

Very few of these antennas have been built to verify their performance, but they have been modeled by a highly experienced Engineer with over 60 years of experience and all of these modeled antennas have great consistencies of results, giving assurance that the modeling process is valid.

There are many individual antennas that have been modeled in various shapes and forms and due to their complexity; many may require trimming in order to achieve the best performance. The conditions of mounting may not be the most desirable and therefore, the best performance may not be achieved. As a result, each and every modeled antenna cannot be guaranteed to perform to the specifications, if you intend to use the data in this book, you must accept it under the conditions that the author is free of any and all liabilities.

First published by Dog Ear Publishing
4011 Vincennes Road
Indianapolis, IN 46268
www.dogearpublishing.net

dog ear PUBLISHING

ISBN: 978-1-4575-4900-7

This book is printed on acid free paper.
Printed in the United States of America

Welcome to My World

I grew up in a time when electronics and radio was in its infancy and most everyone had a great interest in this new and evolving technology. This interest was demonstrated by most of my childhood companions. At the age of 13 and 14, it was common practice in my residential neighborhood in Queens, New York to discuss how radio circuits worked. Often times during games of stickball or baseball we would draw schematic diagrams with chalk on the street pavement and discuss with each other the details of how circuits worked. Some of us would then even build simple receivers. As a young adult, this interest had grown with my Amateur radio activities and US Naval Reserve service. During my service, I devised circuits and antenna placements to overcome heavy intermodulation aboard the "Fifth Amphibious Force Flagship", during World War II.

After the war, during my entire 44 years as a "Senior Development Engineer" and then "Chief Engineer", I developed many products and I maintained a rigorous work schedule. Design problems were taken home, then solved and ready for the next day's business. In doing this, I had created a work ethic that was embedded in my psyche. I have always felt the need to create things that I believed would contribute to others and I have maintained this desire to even this day.

After retirement, at the age of 69, I designed and personally drew the plans for in my opinion, an ideal retirement home and engaged in some of the building of this home. Unfortunately, Hurricane "Andrew" two years later devastated the entire area. Portions of the house had to be rebuilt and once again, I participated in the reconstruction. After living in the home for 10 years, my wife and I decided to move to the mountains of Western, North Carolina. Once again, I drew up the plans; however this time utilizing a CADD software program. At the age of 80, I was still in relatively good health and looking for new projects to conquer.

Our new home is situated in a 116 acre mountain cove, with mountain ridges 3000' feet in height. I was not certain how well my Amateur radio transmissions and reception could be operated from the inside of the bottom of this cove, so I turned to antenna modeling programs in order to provide the best choice. I began modeling a large variety of antennas that might solve my problems; however the modeling became so interesting and captivating that I began to write articles on various designs, culminating in (25) articles for an Online Antenna Magazine. During this period, I began developing problems with macular degeneration in both eyes and I no longer could maintain a suitable schedule.

These articles however encouraged me to continue modeling new antennas and in order to do this I had to obtain computer software that could readily enlarge the monitor display and the monitors were increased in size to 53" inches. The larger monitors helped but in order to read text, a handheld magnifier was utilized.

My typing skills have deteriorated to the point of using a large magnifier and single finger typing. The lack of dexterity in my fingers and the lack of site to view, to what is being typed have brought my touch typing down to at most, hunt and peck. Without some sort of typing aid software, along with speech recognition software as well, typing would be impossible.

The use of converting speech to typing software tends to overcome this problem; however, the accuracy could be improved in the area of speech recognition. Text reading software allows typed text to be read over and over and corrected as necessary. A small video camera with a 22" inch monitor allows me to observe documents with text and or diagrams. This tool allows me some flexibility to model, write text and helped put me back on the way to writing my books.

The large monitors allow me to utilize computer aided design programs necessary for drawing schematics, antenna configurations and charts thereby covering most of the necessary writing activities.

My modeling efforts have led me to write about antenna Broad-Banding and I quickly determined that those antennas that had been previously Broad-Banded and available in antenna hand-books are disjointed from one another and that no overall Unified Broad-Banding message was available. I soon began to model each of the logical possibilities and design them into a "Unified Broad-Banded Method" and write about them. I started designing Broad-Banded antennas in a peculiar way, far different from the way most antennas are Broad-Banded. Instead of using a 1/2wL (wavelength) radiator coupled with its Source to an additional resonant Broad-Banding circuit, instead I at first discovered High-impedance coupling. This took the form of a 10m J-pole Vertical antenna, that was divided into two sections, one of which was the radiator, while the second was the 1/4wL resonant circuit. These are coupled together by a very small value of a Capacitor of just 1 or 2pF.

The original J-pole has a very narrow SWR (Standing Wave Ratio) bandwidth in the 10m band and this was due entirely to the narrow bandwidth of the 1/4wL resonant circuit. After Broad-Banding, this antenna has an SWR of 1.5/1 from 25 all the way to 29.7MHz, which is a bandwidth of 4.9MHz. This change in bandwidth was almost startling to me even after all my years of experience.

Many Broad-Banding configurations were modeled and written about utilizing "Critical-Coupling". Subsequently other forms of Broad-Banding were written about, as was presented in my first book titled, "Super-Dipole Antennas" and then culminated in the "Double Broad-Banding Dual Inverted V Antennas". These antennas are shown and explained in this second book titled, "Super Max-Dipole Antennas".

These two books that I have written have taken quite a few years to complete, to the point of being ready for publishing and the first book has over 640 illustrations, all of which took considerable effort and time to complete. Without the help of Eugene Belton, who inserted the illustrations, reviewed the text and formatted the books, they could not have been completed. Additionally I would like to thank Dale Parfitt, W4OP, President of PAR Electronics, who unselfishly, while spending many of his own hours, reviewed this book for me and offered critical thinking and necessary comments, which I greatly appreciate and cannot thank the two of them enough.

Normally I would have dedicated this book to other individuals or groups; however, after spending years of modeling, building concepts, making unending illustrations and having great difficulties in writing while still having a long rigorous regimen, I have reconsidered. Instead and considering all of the difficulties and limitations, time and effort, I have come to the conclusion that I am most deserving of having it dedicated to myself, Edward J. Shortridge on my 95th birthday.

About the author

Edward J. Shortridge, W4JOQ, has been a licensed Radio Amateur for 76 years. His interest in radio and electronics started early in 1934.

He became a licensed Amateur Radio Operator in 1938. A U. S. Navy radio operator for over 5 years, he served aboard the Fifth Amphibious Group Flagship in landings in North Africa, Italy and Normandy, followed by the Philippines and Okinawa.

As a Chief Electronic Engineer for 44 years, he worked to develop many products at several electronic manufacturing companies. The products were wide ranging, including military and civilian two-way radios, radio repeaters, antennas, ultrasonic communicators, aeronautical beacons, marine depth finders, ground-based satellite receivers and radio equipment for the Moon-landing simulation, in Flagstaff, Arizona. He built and used an Open-Field test range to conduct FCC certification tests for manufactured radio equipment.

As part of one FCC working group, he participated in formulating a portion of the VHF and UHF Land-Mobile radio specification requirements. More recently, since retiring, he has been modeling antennas constantly for over 15 years and has written over 25 articles for an Online Antenna magazine. The content of certain of these articles has led to the development of a "Unified Broad-Banding Method", which forms the foundation for the previous book, titled "Super-Dipole Antennas" and this latest book titled "Super Max-Dipole Antennas".

At the age of 95 and having declining vision problems, finishing this book has been difficult for him to complete.

Preface

A previous book titled "Super-Dipole Antennas", which was written by this author, introduced a completely new Unified Broad-Banding Method. This method primarily consists of the use of a 1/2 wavelength Dipole, which is fed with a 1/4 (or shorter) wavelength resonant line. The amount of coupling between these two components is greatly reduced and controlled to produce a wide SWR bandwidth. This Broad-Banding method is referred to as "Critically-Coupled Broad-Banding".

This new book, titled "Super Max-Dipole Antennas" is a continuation or addition to the previous book and primarily presents Inverted V and Horizontal antennas in the 75-80 and 160m bands, which are some of the widest HF bands.

Early chapters reintroduce the "Unified Broad-Banding Method", presenting Inverted V antennas, with each of the possible "S" Broad-Banding configurations presented at various antenna heights, with a few newer configurations being presented.

A totally different Broad-Banding method, which uses "Dual Inverted V" antennas, which can produce very low SWR bandwidths at all antenna heights, is then presented. Some of these were previously presented but these are new or improved versions.

The latter chapters present some completely new Broad-Banding configurations utilizing "Dual Inverted V" radiators, which in itself creates Broad-Banding but by adding one of several "Critical-Coupling" configurations, "Double Broad-Banding" can be achieved. This arrangement can produce SWR bandwidths at or below an SWR of 1.5/1 across the entire bandwidth at any and all antenna heights. This appears to be a notable achievement, not occurring often.

These new and unique configurations introduced will completely change the perception of how antennas are matched and Broad-Banded. Many of the configurations are very simple and easy to understand. Most have been reduced in complexity, cost of materials and can readily be assembled and these are referred to as "Super-Dipole Antennas".

Throughout this book configurations presented include charts showing how the Far-field RF gains and coaxial losses vary across each of the bandwidths, giving a good perception of the efficiency on each band. All of the antennas throughout this book have been configured with an efficiency in mind and those antennas that have higher losses are scrutinized rigorously and the true performance specifications are shown, whether they are good or bad.

Super Max – Dipole Antennas

Table of Contents

Chapter 1

Chapter 2

Chapter 3

Chapter 3 (cont.)

Chapter 4

Chapter 5

Chapter 6

Super Max-Dipole Antennas

By Edward J. Shortridge W4JOQ

Chapter 1

Basic Super-Dipole Considerations

Introduction

This book has taken the modeling of "Super-Dipole Antennas" to a totally new level with the designs that stretch the imagination. These antennas are unique and have simple configurations that perform well beyond the ordinary.

Both the "Super-Dipole" and "Super Max-Dipole" books have been researched and written by someone with over 13 years of constant modeling antennas experience. Unfortunately Ed who is over 94 years young has declining vision well below being legally blind. Using numerous vision aids allows the design and writing to occur. Extreme care and attention is taken when modeling and writing, some spelling and incorrect wording may occur due to "Voice to Text" translating errors. If you tend to accept minor errors, you will find that the "Super Max-Dipole Antennas" has very interesting information that is presented in an organized manner and to the extent that is not available otherwise.

This book is an extension of the original book titled, "Super-Dipole Antennas", which presented a Unified-Broad-Banding method, which is known as the "Critical-Coupling Method". Large numbers of antenna configurations were devised using this method, resulting in antennas with a significantly extended bandwidth, making them much more suitable to use and eliminating or at least reducing the need to constantly use an Antenna-Tuner as the frequency of operation is altered.

Updated or completely new configurations are included in this new book, most of which improve the SWR bandwidth and match 50 Ohms easier with simpler and easier fabrication methods. This is the first of several requirements to earn the classification of a Super-Dipole Antenna.

The configuration must be capable of readily matching the antenna to a 50 Ohm Source impedance and it must have a means of readily adjusting the amount of coupling between the Radiator and an additional resonant circuit, to the vicinity of "Critical-Coupling". It must be highly efficient, with a minimum of loss if any portion of the added resonant circuit utilizes coaxial cable.

How "Critical-Coupling" is achieved

In order to have Broad-Banding and achieve coupling in the vicinity of critical, there must be two isolated resonant circuits, one of which is the Radiator and the second one can be one of a large variety of configurations. This new book will be mainly concerned with wire antennas in the two widest Amateur bands, namely, 75-80 and 160m. It will strive for low SWR bandwidths, which vary widely with mounting heights. All attempts will be made to achieve matching by the simplest means and the Broad-Banding and matching must be done with a minimum of RF loss in order to achieve the greatest efficiency.

The most basic configuration is a parallel LC circuit, connected in parallel with the Antenna Source. The Inductor can be defined as a wound coil, a length of coaxial cable, or Window/ Ladder-Line. The capacitor can be a discrete component, or replaced with the length of coaxial cable giving a substitute for the capacitive reactants. On the other hand, a 1/4wL resonant line which uses coaxial cable or Window/Ladder-Line as a substitute.

Each of these resonant circuits must be adjusted to present impedance across the Source, which has a value of the Radiator Feed-point impedance. This resonant impedance must be capable of being increased or decreased, giving different degrees of coupling between the Radiator and this resonant circuit. There are many variations to these requirements, some of which give outstanding performance, such as low SWR, high-efficiency, ease of fabrication and adjustment.

Dual Radiators are also used to achieve low and wide SWR bandwidth providing a simpler and different Broad-Banding method. When folded at a downward slope a perfect 50 Ohm match is obtained.

Needs for Broad-Banding

Horizontal wire antennas in the 75-80 and 160m bands cannot achieve low SWR values except for a very narrow bandwidth, making the antenna unusable unless an external Antenna-Tuner is used. If the mismatch of the antenna to 50 Ohms is very high, an Antenna-Tuner can match poor SWR values to the transmitter or transceiver, but the mismatch and poor SWR on the coaxial cable or transmission line is still present and its poor SWR is causing losses, which the Antenna-Tuner cannot remove.

Inverted V antennas in these two bands present an even inferior situation and rotating the Radiators down to a slope of 45° makes things considerably worse. As a result, there have been only a very few Broad-Banding configurations available to purchase and some antenna kit suppliers have not offered them for sale. This is unfortunate, since the Inverted V can be used with a single mast support.

The "Super-Dipole Antenna" has made Inverted V antenna configurations that have overcome this difficulty to a large extent. Nevertheless, the problem is generally overcome with antennas mounted high, but as the antenna is lowered, the SWR once again deteriorates when mounted at 60' feet or lower for use in the 75-80m band. This new book presents antennas that utilize "Double Broad-Banding" as a possible solution, presenting low SWR bandwidths from 100' feet down to 8' feet, as the Radiator slope is altered from 45° down to a horizontal position 8' foot above ground.

Antenna-Tuner considerations

Antenna-Tuners are generally provided in most HF Transceivers and when having an external high-power RF Amplifier, external tuners are used. Most wire antennas in the 75-80m and 160m bands have very poor SWR over the majority of the bandwidth. There is generally a small portion of the bandwidth that might have a satisfactory SWR, but the remainder of the band presents a poor SWR to the transmission line. Depending on the frequency and the length of the transmission line, a wide variety of reactive impedances are therefore presented to the Antenna-Tuner.

The Antenna-Tuner can generally match this impedance to 50 Ohms, providing a good match, so that the full RF power output from the transmitter can be delivered to the transmission line. Unfortunately, the transmission line has remaining standing waves and a poor SWR. In most cases, after the Antenna-Tuner is matched, the average user considers that all is well, but high losses can be taking place in the transmission line, unknowing to the user. The following Inverted V antenna will be examined in order to show this difficulty.

Examining Transmission losses

An Inverted V which is mounted at 60' feet with a Radiator slope of 45° and using #14 AWG wire is shown in Figure 1. It matches 50 Ohms due to the Radiator slope and is a perfect match at only 3.75MHz.

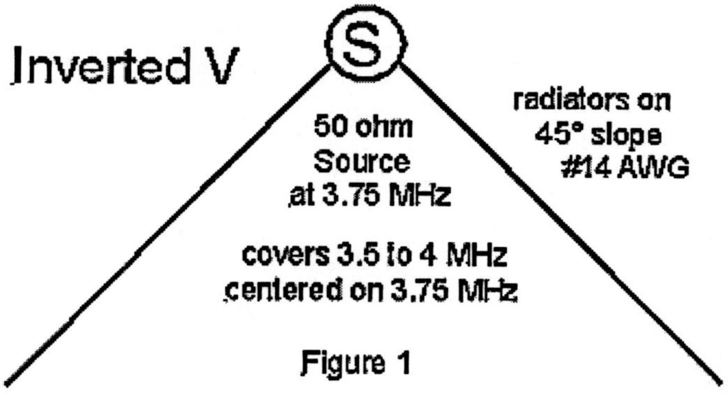

Figure 1

Figure 2 shows the typical equipment connections, with the transmitter connected to the Antenna-Tuner and then connected to the 100' feet of 50 Ohm transmission line. The transmission line is then connected to the antenna Source. The frequency was adjusted to 3.5 MHz, which has the worst SWR of 10/1.

The antenna Impedance at 3.5MHz measured 39 – j 122 Ohms as measured at the Source. When 100' feet of coaxial cable has a specified loss of 0.5dB/100'ft/10MHz, VF = 0.84 is used, it presents an impedance of 210 +j 233 Ohms to the Antenna-Tuner, as shown in Figure 2. The Antenna-Tuner can be adjusted to convert this impedance to 50 Ohms, for which the transmitter is designed.

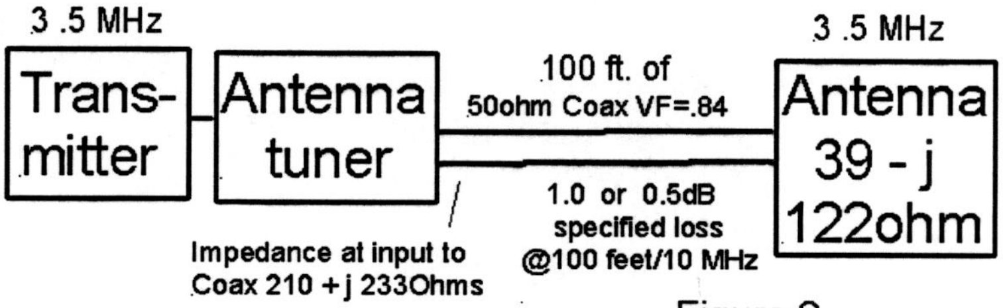

Figure 2

The SWR bandwidth which is shown in Figure 3 has a perfect match at 3.75MHz, with a 10/1 at 3.5MHz and close to 8/1 at 4 MHz. The 1.5/1 SWR bandwidth occupies an extremely narrow usable bandwidth. To be able to use this antenna across the entire 3.5 - 4MHz band, it must be either Broad-Banded or require the use of the Antenna-Tuner.

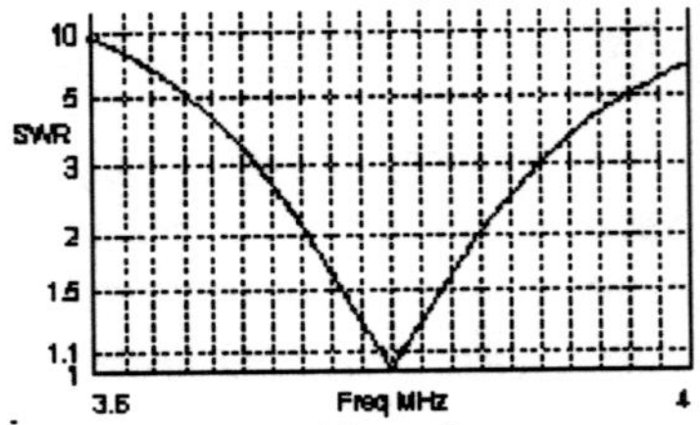

Figure 3

The Far-field RF radiation gain and losses were measured as shown in Figure 4. This is a very important chart which confirms the extent of these losses and the resulting losses are indicated as the difference between the Gain values as shown, in Figures 4, 5 and 6. The gains were measured at 3.75 MHz, which is the center of the 75-80m Amateur radio band. At this frequency the antenna presents a perfect match to 50 Ohms and therefore the only losses should be identical to this specified 100' feet of transmission line.

Far-field Gain & Losses at 53 ft.H All measurements made at 3.75MHz		
Rated Coax loss	Measured loss	
0.00 dB	3.63 dB	
		0.31dB
0.50 dB	3.32 dB	
		0.31dB
1.00 dB	3.01 dB	
Coax loss - 0.5 or 1.0dB/100ft/10MHz		

Figure 4

If the larger diameter coaxial cable, which has a specified loss of 0.5dB, is used, this chart indicates a loss of only 0.31dB as measured at 3.75MHz. The coaxial cable was specified at 10 MHz, but drops to 0.31dB at 3.75MHz and will drop further below that figure at 3.5 MHz. Therefore, for the Far-field measurements, a loss of 0.3dB at 3.5MHz will be used and 0.6dB, when using the smaller diameter coaxial cable, which has a specified loss of 1.0dB.

In order to determine the radiation loss at 3.5MHz, the chart as shown in Figure 5 will be used. At this frequency the Inverted V has an SWR of 10/1. This shows a loss of 1.41dB when using coaxial cable with a specified loss of 0.5dB and 3.49dB when using a specified cable loss of 1.0db. This is a substantial loss of power that is still present after an Antenna-Tuner has been used and there will be no obvious indication that it is occurring. If you are using high RF power of 1000W or there about, you could easily be dissipating 250W of power in the transmission line cable, this could easily damage the coaxial cable and cause a malfunction.

Far-field Gain & Losses at 53 ft.H All measurements made at 3.5MHz		
Rated Coax loss	Measured loss	
0.00 dB	3.46 dB	
		1.41dB
0.50 dB	2.05 dB	
		2.08dB
1.00 dB	0.97 dB	
Coax loss - 0.5 or 1.0dB/100ft/10MHz		

Figure 5

If the antenna was adequately Broad-Banded to an SWR of approximately 2/1 and an Antenna-Tuner is used, these losses would be greatly decreased.

The SWR chart for 4 MHz, as shown in Figure 3 indicates that the Inverted V antenna has an SWR in the vicinity of 7/1 to 8/1 will have losses less than those at 3.5MHz. At this end of the bandwidth, the loss has dropped to 1.00dB when using the larger coaxial cable and 1.83dB when using the smaller diameter coax with the specified loss of 1.0dB/100'ft/10MHz as shown in Figure 6.

Far-field Gain & Losses at 53 ft.H All measurements made at 4.0MHz		
Rated Coax loss	Measured loss	
0.00 dB	3.74 dB	
		1.00dB
0.50 dB	2.74 dB	
		0.83dB
1.00 dB	1.91 dB	
Coax loss - 0.5 or 1.0dB/100ft/10MHz		

Figure 6

These charts and measurements have confirmed that there are high RF power losses in long coaxial transmission lines, when measured at the band edges where the SWR is quite high. These losses are hidden, with no obvious means of recognition to the user. The Inverted V antenna presents a high SWR at the band edges, but the Antenna-Tuner cannot help the mismatch unless the transmission line is removed. If you use a remote Antenna-Tuner, it can present a perfect match to a 50 Ohm transmission line leading to the transmitter. Under these conditions, there will not be extra losses in the transmission line beyond the basic specified losses, which cannot in any way be removed.

Basic Inverted V Antenna configurations

The original "Super-Dipole Antennas" book covered a wide variety of Broad-Banded antenna configurations, all of which fit into the "Critical-Coupling" unified Broad-Banding system. This book, the "Super Max-Dipole Antennas" will review many of the Inverted V antennas, presenting slightly different and improved configurations. Then, newer configurations that have reduced losses and better efficiency will be shown and discussed.

Center-fed "Sc" Inverted V Antennas

This antenna is an Inverted V antenna that is fed with a number 1/4wL shorted line, tapped part way down to a 50 Ohm Source. In order to achieve any semblance of Broad-Banding, this line must have a characteristic impedance of approximately 12.5 Ohms. If this line impedance is 50 Ohms, <u>Absolutely no Broad-Banding can occur.</u>

There does not seem to be any practical way to obtain a coaxial cable with this low of impedance, such as 12.5 Ohms, but it can be modeled and show significant Broad-Banding, otherwise it is only of academic interest and therefore this antenna is not shown.

Center-fed "St" Inverted V Antennas

Means can be taken to transform a 50 Ohm line so that it appears to be 12.5 Ohms. A Balun step-up transformer can be used to step up the impedance at the Feed-point from Wire 2 to approximately 200 Ohms. Then, a 1/4wL 50 Ohm line that is shorted at the bottom end and tapped-up to a 50 Ohm Source can be used. Such an arrangement can be seen in Figure 7. In doing this, it is called a Transformer-enhanced Broad-Banding line, or a "St" line. The 50/200 Ohm Balun is being used in a reverse direction from its normal usage. In this case, the 50 Ohm end of the Balun is connected to Wire 2 and the 200 Ohm then is connected to the 1/4wL line. The standard 4/1 Balun generally uses approximately 22" inches of Twin Wire, having an impedance of approximately 100 Ohms. This is a relatively short length compared to the rest of the 1/4wL line lengths, but it must be taken into account since it has not been included into the modeling of this antenna.

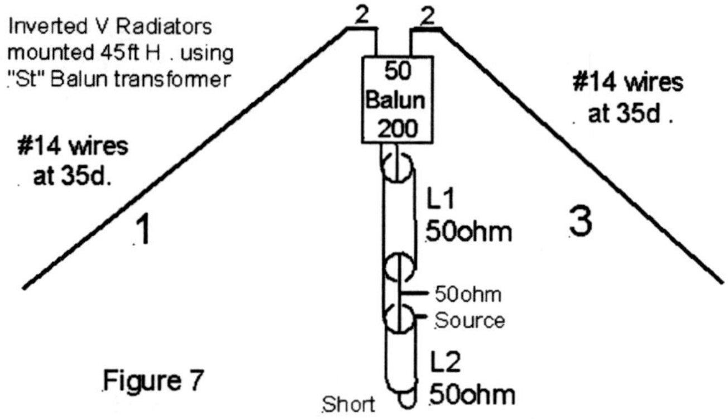

Figure 7

The 50/200 Ohm Balun can be considered balanced at the 50 Ohm end. Remember, a Balun is a four terminal network and whether one end or the other is balanced depends on what terminal is intended to be considered ground. In this particular case, one terminal of the 200 Ohm end is intended to be connected to the 1/4wL coaxial line braid and the low impedance side of the Balun is not connected to any coaxial cable and is every bit balanced as you might expect.

This antenna was modeled at a height of 45' feet above ground, as shown in Figure 7. At this height it is difficult to achieve good Broad-Banding, but it was chosen because of this difficulty and additionally many Amateur radio operators cannot obtain Mast heights much greater. At this height the radiators could not be at a slope of 45° without the Radiator ends being too close to ground, so therefore an angle of 35° was chosen. At this slope, the Radiator ends are well above ground at 9.1' feet, out of reach.

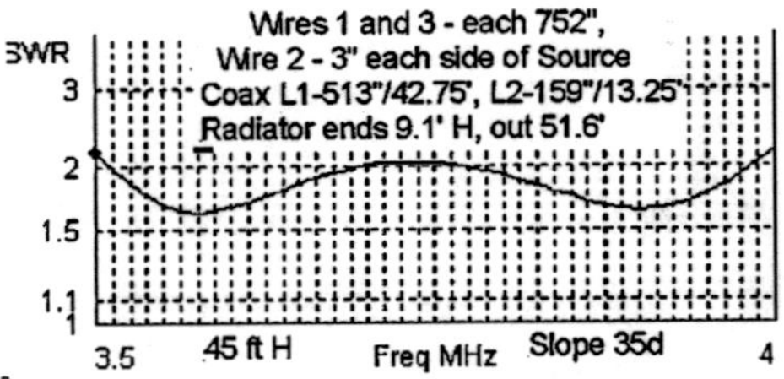

Figure 8

The antenna was scanned across the 75-80m band and the resulting SWR plot as shown in Figure 8 has an SWR averaging better than 2/1. The best SWR was achieved when the Broad-Banding line lengths for L1 and L2, were 513" inches and 159" inches respectively, as shown in Figure 8. At these line lengths, the best Source impedance of 50 Ohms was achieved.

Far-field Gain & Losses at 90d.			
45ft.H	3.5	3.75	4.0MHz
No loss	3.60	3.84	4.03dBi
w/ loss	1.54	3.42	2.58dBi
RF Loss	2.06	0.42	1.45dB
Coax loss - 0.5dB/100ft/10MHz			

Figure 9

Looking at the Far-field RF Gain and Losses chart of Figure 9 tells a different story. The RF loss at 3.5 MHz was 2.06dB and 1.45dB at 4MHz. These SWR figures are not very good and the only advantage is that you could operate this antenna primarily with low-power RF levels without the use of an Antenna-Tuner. Otherwise, the actual losses are as bad as, or worse than, those shown in Figure 5.

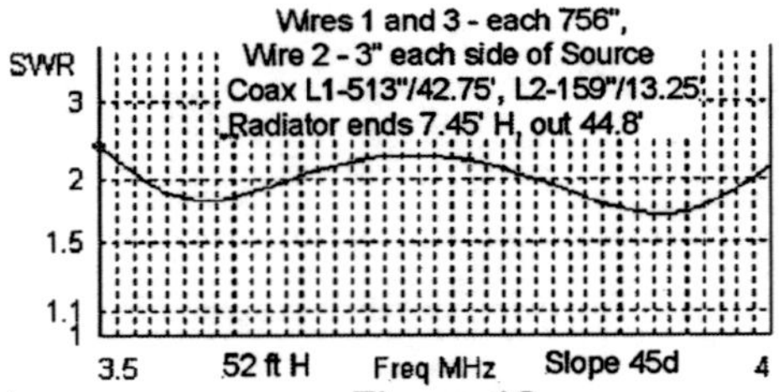

Figure 10

The RF gain figures throughout the entire band are very good but not achievable unless the coaxial cable losses are removed. This antenna's Broad-Banding plot looks good, but it certainly is not an efficient antenna.

The antenna was raised to 52' feet so that the Radiator slope of 45° can be used. In doing this the ends of the radiators are at 7.45' feet above ground. With these changes, the antenna was scanned for SWR response and the results are shown in Figure 10. Once again, the SWR plot looks quite good but not quite as good as shown in Figure 8.

Far-field Gain & Losses at 90d.			
52ft.H	3.5	3.75	4.0MHz
No loss	3.43	360	3.72dBi
w/ loss	1.09	3.21	2.29dBi
RF Loss	2.34	0.39	1.43dB
Coax loss - 0.5dB/100ft/10MHz			

Figure11

Due to the additional slope to 45°, the Far-field RF radiation Gain and Losses chart of Figure 11 has deteriorated further, with losses of 2.34 and 1.43dB at 3.5 and 4MHz respectively.

This antenna clearly demonstrates that good SWR values can be obtained by this configuration of Broad-Banding but when there are long coaxial lines used to achieve these low SWR values, care must be used to make certain that the configuration is efficient.

Coaxial cable with its inherent losses, must not be used in configurations where high voltages and currents are built up on this cable, otherwise these coaxial lines will have excessive losses. This gives an incentive to look for configurations to avoid such losses. Then antennas such as this one would be very good and would perform very well.

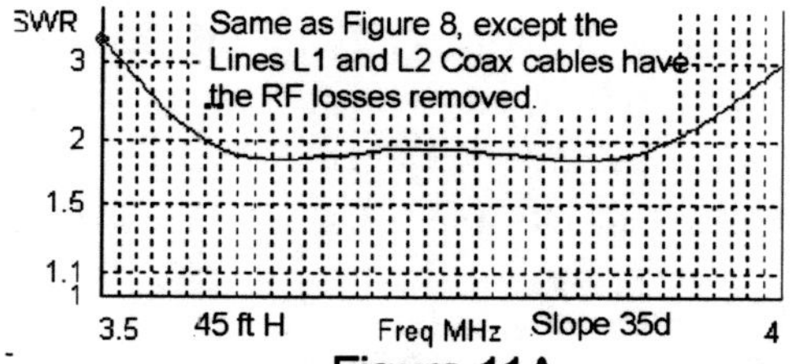

Figure 11A

Many Broad-Band antennas, such as this one, may achieve a good portion of its Broad-Banding due to the resistive losses in the coaxial cable. This antenna was checked to determine what the effects on the SWR bandwidth plot might be. The L1 and L2 coaxial losses were removed in the modeling process and a new SWR plot Figure 11A was made. The SWR plot was still satisfactory in the center portion of the bandwidth but it was slightly over 3/1 at the band edges. Removing the losses certainly affected the SWR response and it was detrimental but the losses in SWR were considerably less than expected.

Off-Center "So" Inverted V Antennas

In the previous "St", the Balun-enhanced method was used to increase the impedance to a level of approximately 200 Ohms, so that Broad-Banding could be achieved, while using a 50 Ohm coaxial cable. Another method is to dispense with the Balun and feed the antenna off-center to an impedance of approximately 200 Ohms. This is referred to as a "So" configuration and is shown in Figure 12.

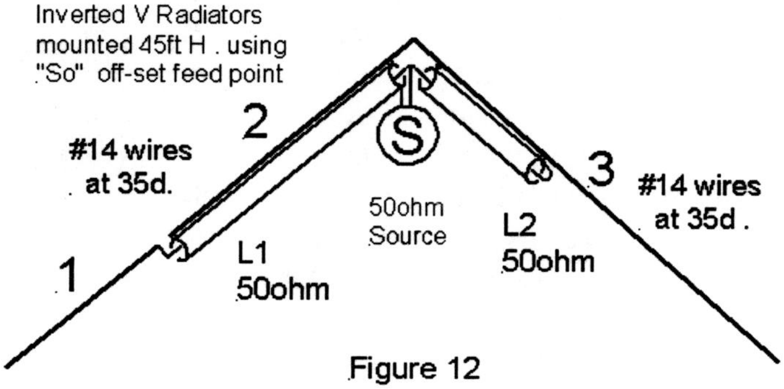

Figure 12

This antenna uses #14 AWG wire and is mounted at a height of 45' feet with a Radiator slope of 35°. The Feed-point is located 252.6" inches from the end of Wire 1 and a 50 Ohm coaxial cable is used for a 1/4wL resonant Broad-Banding line that is shorted on the far end.

A 50 Ohm source is tapped up about 30% from the shorted end of the cable, which consists of L1 and L2. It is very important in the design to make certain that the length of the coaxial cable L1 is exactly the same length as Wire 2 so that the junction of L1 and L2 are precisely at the Radiator Apex of the Inverted V and in this case both of the lengths are exactly 502" inches.

Coaxial cable with a loss of 0.5dB/100'ft/10MHz was used for the lines L1 and L2. The antenna was scanned over the 75-80m band and the results can be seen in Figure 13. The resulting SWR response of 2/1 averages approximately the same as the previous "St" antenna, both of which are mounted at 45' feet with an antenna slope of 35°.

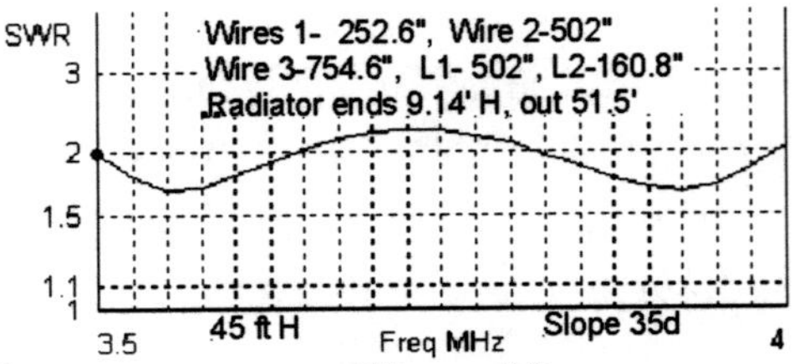

Figure 13

Far-field Gain & Losses at 89.7d.			
45ft.H	3.5	3.75	4.0MHz
No loss	3.46	3.83	4.11dBi
w/ loss	1.60	3.44	2.77dBi
RF Loss	1.86	0.39	1.34dB
Coax loss - 0.5dB/100ft/10MHz			

Figure 14

The Far-field gain and losses chart of Figure 14 also shows high RF radiation losses at both ends of the 75-80m band, but this antenna has slightly less loss, but it is still excessive. If coaxial cable with the specified loss of 1.0dB was used, the losses would be almost doubled to totally impractical values.

Lowering the antenna to 25' feet (as shown in Figure 15) will cause the radiators to touch ground. Therefore, the Radiator slope was changed to 15°, placing the radiator ends at 8.93 ft. above ground. The resulting SWR response was almost identical to having it at 45' feet of height, as shown in Figure 13.

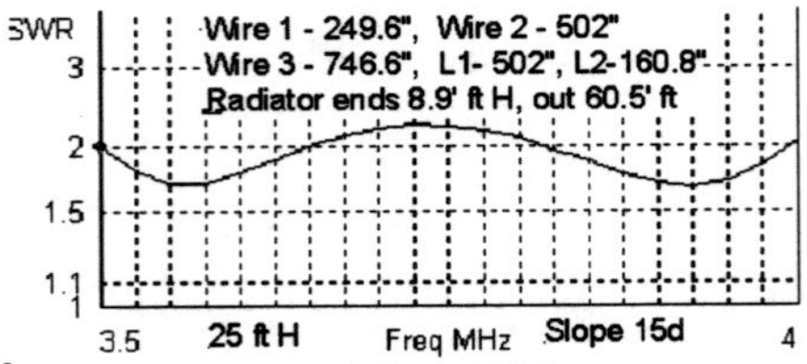

Figure 15

The RF losses, as shown in Figure 16 are substantially the same as of the previous height. Although these losses are considered high, there are Broad-Band antennas that are currently being used with losses as great and the users are unaware that these losses exist.

These two antennas, namely the "St" and "So" were purposely shown first, which has been the normal progression covered in the Critically-Coupled Unified Broad-Banding Method. The previous book "Super-Dipole Antennas" showed other configurations which also had high RF losses in the coaxial cables. It should be quite obvious that if coaxial cables are to be used, attempts should be made to limit it to strictly transmission line use. There are plenty of situations where a short length of coaxial cable can readily be used, but care must be exercised.

Far-field Gain & Losses at 89.7d.			
25ft.H	3.5	3.75	4.0MHz
No loss	1.97	2.49	2.93dBi
w/ loss	0.09	2.11	1.60dBi
RF Loss	1.88	0.38	1.33dB
Coax loss - 0.5dB/100ft/10MHz			

Figure 16

Short-line "Sx" 45° Inverted V Antennas

The very basic "S" Broad-Banding configuration, consisted of discrete components of a capacitor and inductor combination. This "Sx" configuration still uses a discrete capacitor but then an inductor is replaced by a short section of 450 Ohm Window/Ladder-Line. The values of these two items are chosen so that they resonate in the vicinity of 3.75 MHz. The impedance of this resonant circuit is varied by changing the capacitor value and the Window/Ladder-Line length, while still maintaining the same resonant frequency.

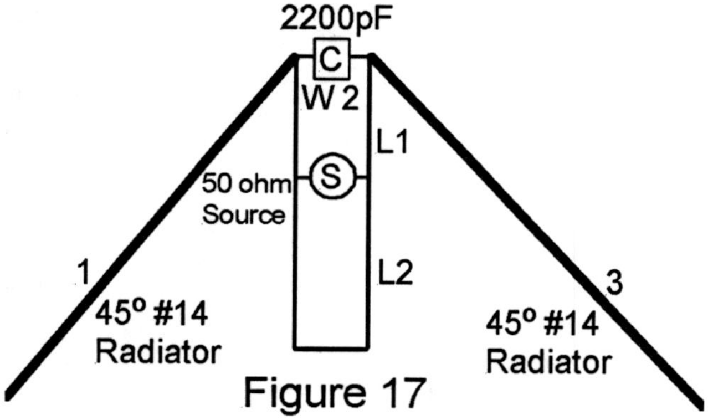

Figure 17

The Inverted V, which has a Radiator slope of 45°, normally has Source impedance at Wire 2 (W2) that is approximately 50 Ohms but when this "Sx" resonant circuit is used, the impedance at Wire 2 (W2) will be higher than the normal 50 Ohms and therefore the Source must be tapped-down on this line as shown in Figure 17.

When this antenna is mounted at 85' feet above ground and the 2200pF capacitor is used, lines L1 and L2 are set at 6.8" inches and 12.8" inches respectively. If the capacitor tolerance is very good, it will resonate close to 3.75MHz and create an SWR response as shown in Figure 18. The resulting SWR bandwidth averages at approximately 2/1 across the 75-80m bandwidth.

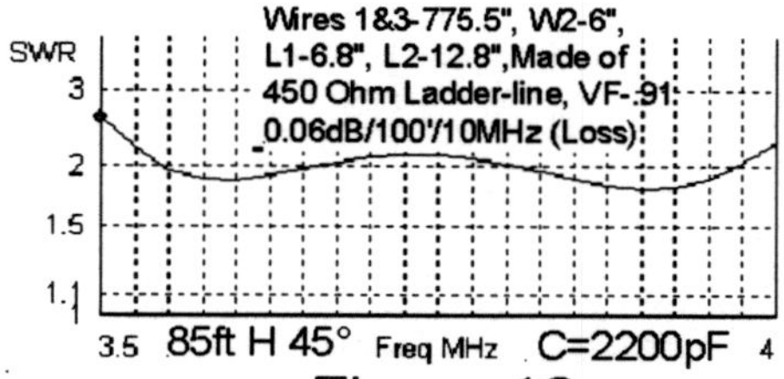

Figure 18

To allow for the Capacitor tolerances, it would be wise to extend line L2 about 2" inches longer and provide an adjustable clamp that can be slipped up or down to extend or reduce the L2 line length.

Far-field Gain & Losses at 89.5 d.			
85ft.H	3.50	3.75	4.0MHz
No loss	5.03	5.00	5.00dBi
With loss	4.78	4.96	4.82dBi
RF Loss	0.25	0.04	0.18 dB
Ladder-line loss 0.06dB/100'/10MHz			

Figure 19

This is a very good SWR when considering that the Antenna slope is at 45°, which generally restricts the bandwidth. The Far-field gain and losses chart of Figure 19 indicates very good RF gains and the RF loss in the Window/Ladder-Line has been reduced substantially to levels of 0.25, 0.04 and 0.18dB across the entire bandwidth. This is a substantial improvement over those losses, as shown in the two previous antennas, both of which used 50 Ohm coaxial cable. The loss in the Window/Ladder-Line is specified at 0.06dB/100'ft/10MHz and the total length of L1 plus L2 is approximately 19.6" inches in length, rather than being a full 1/4wL. The same Inverted V, which slopes at 45°, is used and only the Broad-Banding has been changed to the "Sx" configuration.

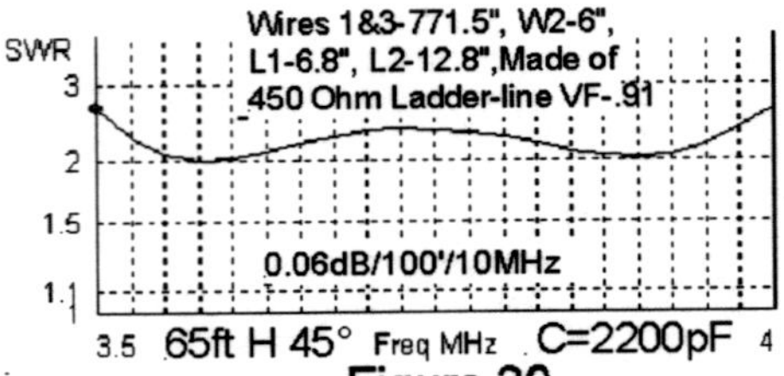

Figure 20

This antenna was lowered 20' feet to a height of 65' feet above ground and in doing this the radiator lengths had to be decreased from 775.5" inches down to 771.5" inches in order to once again have the SWR centered. Surprisingly, the SWR has deteriorated only a very minor amount from the previous height as shown in Figure 20.

Due to the lower mounted height, the Far-field RF gains have dropped slightly and the RF losses only change in a very minor way, as can be seen in Figure 21.

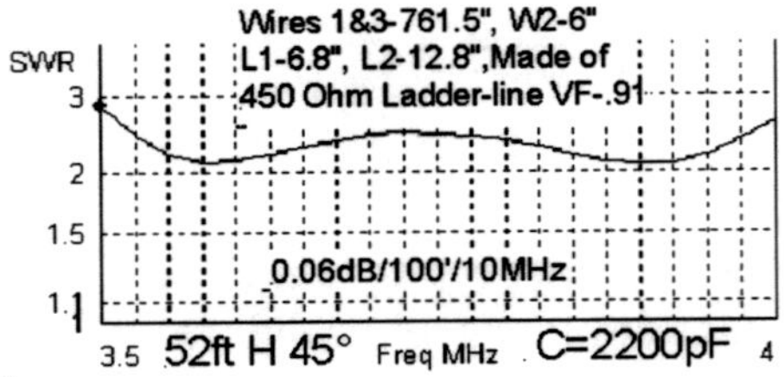

Figure 22

Far-field Gain & Losses at 89.5 d.			
65ft.H	3.50	3.75	4.0MHz
No loss	4.68	4.70	4.68dBi
With loss	4.43	4.66	4.42Bi
RF Loss	0.25	0.04	0.26dB
Ladder-line loss 0.06dB/100'/10MHz			

Figure 21

When mounted at 52' feet which are about as low as possible and still keeping the radiator ends out of reach, they will be down to 7.13' feet above ground and reach out to a distance of 45.13' feet.

The SWR has deteriorated slightly and averages lower than 3/1 across the entire 75-80m bandwidth as shown in Figure 22.

The Far-field RF radiation gains have dropped slightly over 1dB across the bandwidth, as might be expected, the RF losses for the Window/Ladder-Line have not changed and are low, as shown in Figure 23.

This antenna mounted between 52' and 85' feet or more, should work extremely well. It has a good SWR, low losses, can be used with a 50/50 Ohm Balun and is very efficient and capable of handling high RF power. The voltages that might appear across the 2200pF capacitor will only be slightly higher than those supplied to the Source.

Far-field Gain & Losses at 89.6 d.			
52ft.H	3.50	3.75	4.0MHz
No loss	3.38	3.55	3.67dBi
With loss	3.10	3.51	3.46dBi
RF Loss	0.28	0.04	0.21 dB
Ladder-line loss 0.06dB/100'/10MHz			

Figure 23

By adjusting the Source point on the L1/L2 resonant line, the SWR response can be altered to some degree. All of these combined, puts this antenna into a class called a "Super-Dipole Antenna".

Short-line "Sx" 35° Inverted V Antennas

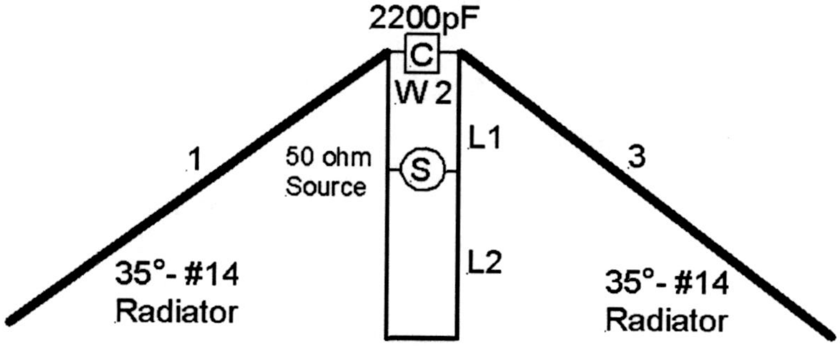

Figure 24

This antenna has the same configuration as the previous antenna except that the radiators have a 35° slope, as shown in Figure 24. By having less radiator slope, the SWR bandwidth should improve and possibly the RF losses may decrease.

At a height of 85' feet, the SWR has notably improved averaging below 2/1. The radiator wire lengths had to be altered, along with the lengths of lines L1 and L2 to values of 7.0" and 12.8" inches respectively, as shown in Figure 25.

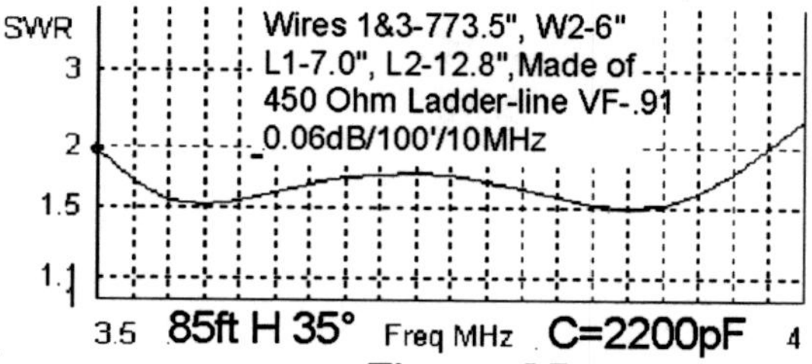

Figure 25

The Far-field gain and loss chart of Figure 26 shows improvement in both the RF gain and a reduction in the antenna losses. These improvements would continue further if the radiator slope is reduced to a horizontal position.

Lowering the antenna down to 65' feet and then to 45' feet above ground, the SWR deteriorates slightly, as shown in Figures 27 and 29. At 45' feet above ground the SWR averages slightly over 2/1, SWR, which is still very good.

Far-field Gain & Losses at 54 d.			
85ft.H	3.50	3.75	4.0MHz
No loss	5.40	5.39	5.42dBi
With loss	5.21	5.34	5.24dBi
RF Loss	0.19	0.05	0.18 dB
Ladder-line loss 0.06dB/100'/10MHz			

Figure 26

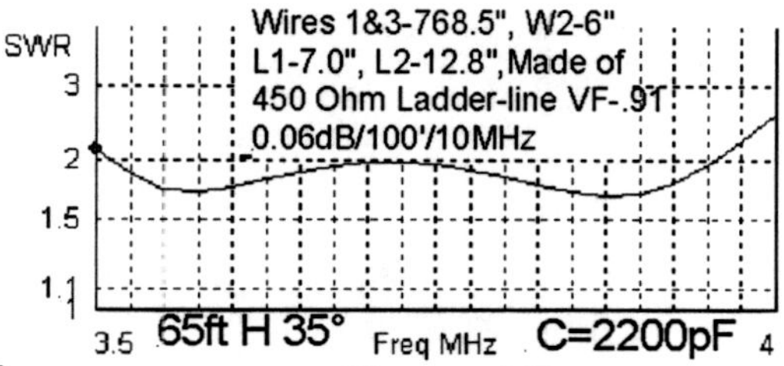

Figure 27

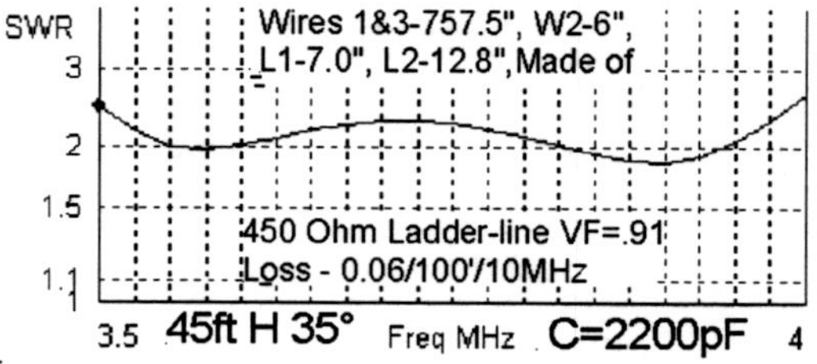

Figure 29

As the antenna is lowered from 85' feet to 65' feet there is only a very minor reduction in the RF gain across the entire bandwidth but the RF losses have increased very slightly as shown in Figure 28.

Far-field Gain & Losses at 89.6 d.			
65ft.H	3.50	3.75	4.0MHz
No loss	5.34	5.33	5.30dBi
With loss	5.13	5.28	5.10dBi
RF Loss	0.21	0.05	0.20 dB
Ladder-line loss 0.06dB/100'/10MHz			

Figure 28

When the height is reduced from 65' feet to 45' feet, a much greater decrease in Far-field RF radiation gain has occurred however it seems typical for that height. Figure 30 shows this drop along with some further increase in antenna loss but these losses are very low, when comparing them with the previous antennas that used 1/4wL resonant Broad-Banding lines.

Far-field Gain & Losses at 83.7 d.			
45ft.H	3.50	3.75	4.0MHz
No loss	3.56	3.79	3.98dBi
With loss	3.32	3.75	3.78dBi
RF Loss	0.24	0.04	0.20dB
Ladder-line loss 0.06dB/100'/10MHz.			
Radiator 35d ends at a 8.79' hieght.			

Figure 30

The "Sx" is a simple configuration that could readily be fabricated and it has extremely high performance, using a capacitor that is resonated with a short length of 450 Ohm Window/Ladder-Line, which is readily available and fairly easy to alter its length. Where the antenna radiator slope is used at a 45° or 35° degree angle, it is readily adjustable and performs in an excellent manner.

Short-line "Sg" 35° Inverted V Antennas

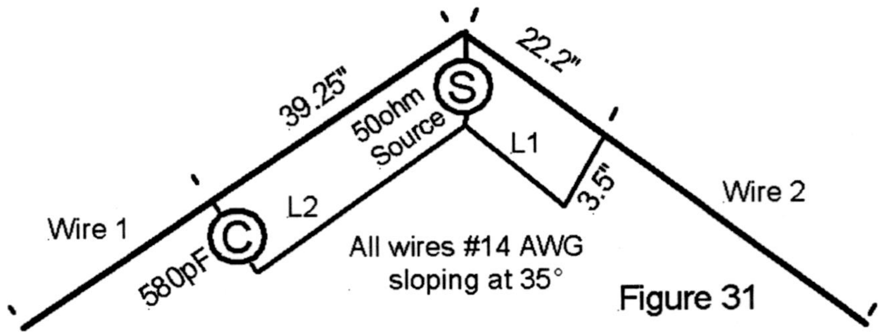

Figure 31

This "Sg" antenna, as shown in Figure 31, is a Gamma-match, configured to be compatible with an Inverted V antenna with a radiator slope of 35°. There are very good 3.5" inch transmission line insulated spacers that are readily available and can be used to separate and hold the wires in place.

In building a Gamma-match to a 50 Ohm Source, it has been the practice to select arbitrary wire lengths, wire spacing and capacitor values as long as it produces a Source of 50 Ohms, which is in series with the capacitor. Most of the time, these values were selected to be convenient in the antenna design.

Unfortunately, the Gamma-match is over-coupled to the radiator, producing 2 resonant frequencies, one of which is used within the band of interest, while the other generally is far removed outside of the typical bandwidth. This second resonance is overlooked and the over-coupling is not obvious. The Gamma resonance within the bandwidth generally meets the intended requirement of matching, but its Broad-Banding capabilities are totally overlooked.

If the resonance that is outside of the intended bandwidth can be moved into the intended bandwidth and properly adjusted, the SWR bandwidth can be greatly improved. This "Sg" antenna will demonstrate how that can be obtained.

Normally, the Gamma-match has been used for the much higher frequency bands, particularly when using aluminum tubing for both the radiator and the matching. In the 75-80m band a wire Gamma-match can be configured using relatively short wire lengths, such as, 39.25" inches and 22" inches for lines L2 and L1. Different line lengths and capacitor values have been used to determine the best combination and the use of a 580pF capacitor with these stated line lengths seems to produce the best SWR bandwidth.

When this combination is used, if the Source is placed in parallel with the capacitor, the impedance would be greater than 50 Ohms and it would be necessary to Tap-down to a 50 Ohm point. This Source was placed at the Apex of the Inverted-V. It was intended that the braid of the transmission line coaxial cable be connected to the radiator Apex if a Balun is not intended to be used, while the center-conductor is connected to the lower end of the Source. Nevertheless, a 50/50 Balun can be used without regard to this polarization.

This antenna was mounted at 100' feet above Average-soil and adjusted for the best SWR bandwidth, which is shown in Figure 32. The SWR averages slightly above 1.5/1 over most of the bandwidth and rises to 2/1 at the band edges.

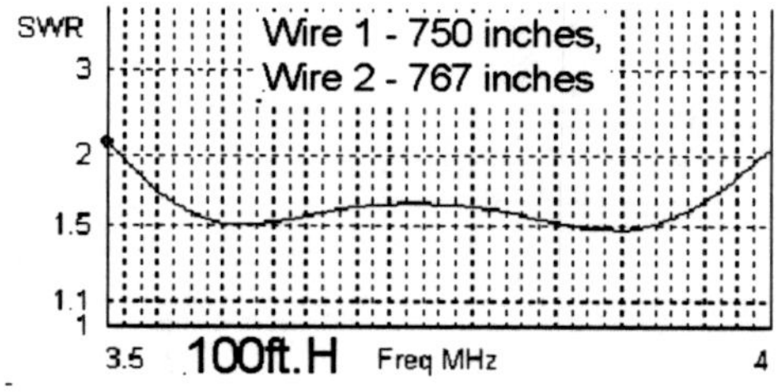

Figure 32

The 580pF capacitor should be substantially lossless, since many well designed capacitors have Q values up to approximately 5000, therefore there is nothing in this antenna to cause RF loss except for the #14 AWG wire resistance. Then when using a capacitor of 580pF the Far-field radiation gain and loss chart of Figure 33 shows very good gains.

Far-field Gain & Losses at 43.3 d.			
100'H	3.50	3.75	4 MHz
With Cap.	5.24	5.58	5.62dBi
With Stub	5.08	5.52	5.39dBi
Stub loss	0.16	0.06	0.23dB
Capacitor=580 pF, Stub=21ft long.			
Coax loss - 0.5dB/100 feet/10 MHz			

Figure 33

A Short-line Open-ended 50 Ohm coaxial cable with a length of 21' feet can be used in place of the 580pF capacitor and the same SWR bandwidth can be obtained when using coaxial cable with a specified loss of 0.5dB/100'ft/10MHz. When the Far-field RF radiation gain was measured with this 21' foot long Stub. The difference in RF gain between "With Cap" and "With Stub" is very little such as 0.16, 0.06 and 0.23dB across the entire 75-80m bandwidth.

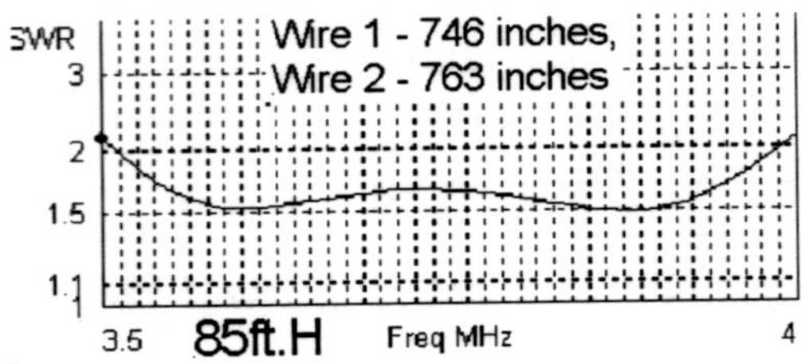

Figure 34

The antenna was lowered from 100' feet, down to 85' feet above ground and surprisingly the SWR has remained averaging just slightly above 1.5/1, as shown in Figure 34.

The Far-field RF radiation gains as shown in Figure 35 have decreased very little and the RF losses when using a Coaxial Stub are substantially unchanged.

Far-field Gain & Losses at 54.4 d.			
85ft.H	3.50	3.75	4 MHz
With Cap.	5.10	5.33	5.25dBi
With Stub	4.94	5.27	5.01dBi
Stub loss	0.16	0.06	0.24dB
Capacitor=580 pF, Stub=21ft long.			
Coax loss - 0.5dB/100 feet/10 MHz			

Figure 35

Reducing the antenna height down to 65' feet causes some further SWR deterioration, as shown in Figure 36 but it is still well below 2/1. Once again, the RF losses using a Coaxial Stub have remained relatively the same, as shown in Figure 37.

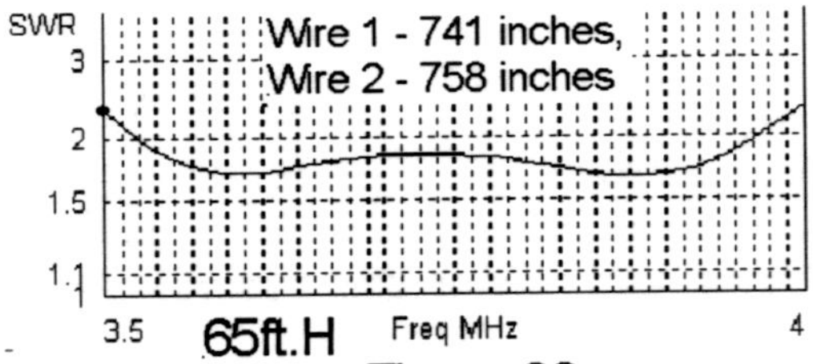

Figure 36

Far-field Gain & Losses at 89.4d.			
65ft.H	3.50	3.75	4 MHz
With Cap.	5.00	5.27	5.09dBi
With Stub	4.83	5.22	4.84dBi
Stub loss	0.17	0.05	0.25dB
Capacitor=580 pF, Stub=21ft long.			
Coax loss - 0.5dB/100 feet/10 MHz			

Figure 37

Lowering the antenna further down to 45' feet might be more practical for most Amateur radio operators to install, and the SWR has decreased further averaging very close to 2/1, which is considered a good practical level to operate with as shown in Figure 38.

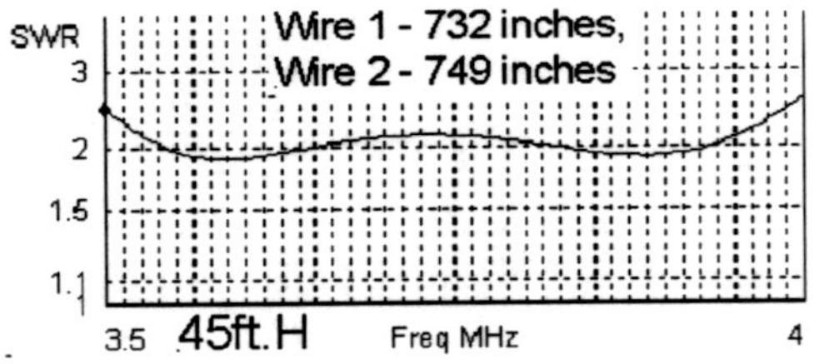

Figure 38

Far-field Gain & Losses at 89.6d.			
45ft.H	3.50	3.75	4 MHz
With Cap.	3.13	3.66	3.67dBi
With Stub	2.96	3.61	3.38Bi
Stub loss	0.17	0.05	0.29dB
Capacitor=580 pF, Stub=21ft long.			
Coax loss - 0.5dB/100 feet/10 MHz			

Figure 39

The Far-field RF radiation gains have dropped some, but are still very good across the entire bandwidth. When using a Coaxial Stub the RF Losses Are still extremely low and consistent with the previous charts, as shown in Figure 39.

Whether you use a 580 pF capacitor or a Coaxial Stub, this antenna provides substantially the same SWR values at all of the various heights above ground and when using a Short-Line Stub all of the RF loss values are consistent. The common wires, which consist of lengths of 39.25" inches and 22.2" inches, control the amount of RF coupling between the radiator and the "Sg" Gamma-match. This antenna provides good SWR values and the plot can be altered by adjusting this coupling. It has good gains, extremely low RF losses and it is easy and inexpensive to construct. With these characteristics it should be considered one of the "Super-Dipole Antennas".

How the "Coupling" affects the SWR

In the previous book titled, "Super-Dipole Antennas", a statement was made that needs further clarification. The statement, "Less is More", refers to the SWR Broad-Banding that is obtained for different amounts of coupling between the radiator and a resonant line or circuit.

In the case of this "Sg" Gamma-match, invariably the common coupling line between the radiator and the resonant Broad-Banding line or circuit is selected far longer than necessary to achieve good SWR Broad-Banding.

In other words, these two items are generally over-coupled. If the coupling is reduced (Less), the two existing resonant frequencies become closer together. When the coupling is reduced adequately, increased (More) in-band Broad-Banding is achieved. Therefore, "Less is More".

As the second resonant frequency is moved into the desired bandwidth, generally a point is reached where the coupling can be considered close to "Critical-Coupling". At this point small adjustments to the coupling and the Resonant circuit or Resonant line frequency can be made, to achieve the very best and the lowest SWR across the entire bandwidth.

Other configurations, such as the "Sx" antennas do not have long coupling lines between the radiator and the resonant Broad-Banding line.

Such antennas are basically tightly coupled, but the coupling is reduced to "Critical-Coupling" by using a low-impedance resonant line and large capacitor that has very low impedance. This will thereby have the equivalent of an extremely short coupling-line.

There is a limit to reducing the coupling below the "Critical-Coupling" point, at which the Broad-Banding begins to disappear and a single resonant point will be obtained within the desired band.

On the other hand, having excess coupling while trying to achieve good in-band, Broad-Banding may be used to an advantage. If this coupling is greatly increased, it is possible to adjust this coupling so that an antenna can actually cover 2 Amateur radio bands. When this is done, there will be no in-band Broad-Banding in either of these two bands.

The previous book titled "Super-Dipole Antennas" presented quite a few Dual-Band antenna configurations in the 20m to 10m bands. Generally, the SWR was adequate since most of these bands are relatively narrow in frequency.

Short-line "Sd" Super-Delta Antenna

The Super-Delta antenna was first introduced by this author in an article in the online AntenneX magazine and very many expanded versions were included in the previous "Super-Dipole" book.

This is a relatively simple antenna to understand and build and it has some of the lowest coaxial cable losses that you might measure. This antenna, as shown in Figure 40, is an Inverted V using #14 AWG wire and has its radiators sloping at 45°.

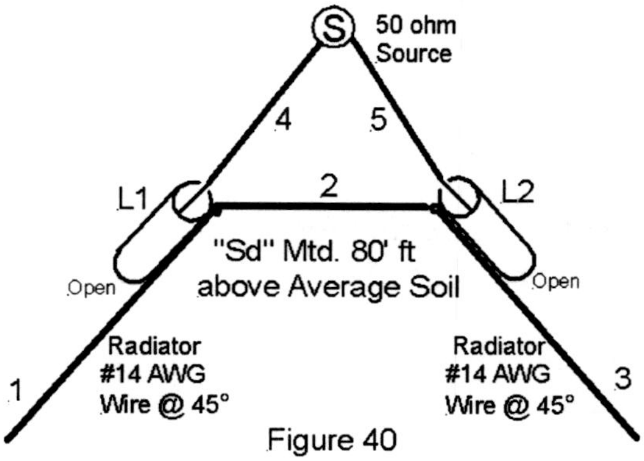

Figure 40

It is shown mounted at 80' feet above average soil. The radiator consists of Wires 1, 2 and 3. With the addition of Wires 4 and 5, a Delta shaped configuration is formed with a 50 Ohm Source located at the very top of the Delta.

This Delta, which consists of Wires 2, 4 and 5 is resonated to the center of the 75-80m band by the addition of 2 capacitive Open-ended coaxial Stub lines L1 and L2. In order to give this antenna the capabilities of operating at the maximum legal power in this band, RG-213 solid-dielectric cable was used. This cable which has a specified loss of 0.6dB/100'ft/10MHz has a velocity factor of 0.66 and because of the solid dielectric; the RF breakdown voltage is extremely high. Additionally the 2 coaxial Stubs L1 and L2 are in series, thereby reducing the RF voltage to one half, giving an extremely high voltage rating.

Since these two open-ended coaxial lines are in series, there are extremely low RF losses and the full RF power is radiated. At a mounting height of 80' and 60' feet, each capacitive stub has a length of 128.5" inches, which is only 10.708' feet long and mounting it along side of the radiator wires should be a relatively simple task.

Even though Capacitive Stubs are used to resonate the Delta Wires 2, 4 and 5, Fixed capacitors with a value of approximately 330pF can be used to replace each capacitive Stub but due to the capacitor tolerance, the SWR plot might be skewed somewhat and require some adjustment. To do this, a small amount of capacity must be added or removed in order to have a centered display. The second choice, is to change the lower lengths of Wires 4 and 5, making them slightly shorter or longer but both at the same length.

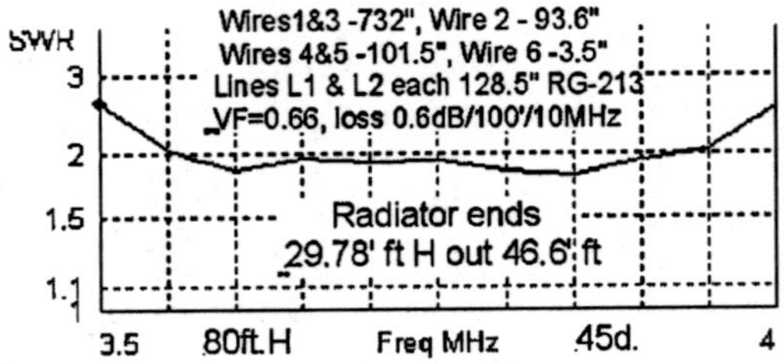

Figure 41

When using the Fixed capacitors all of the SWR plots will be unchanged and the Far-field RF gain will increase by a very small amount. When this antenna is mounted at a height of 80' feet, the SWR plot of Figure 41 shows that the SWR averages slightly lower than 2/1 across the center bandwidth and approximately 2.7 at the band edges. Considering that this antenna has the radiators sloping at 45° and the SWR value of 2/1 seems like a suitable value to obtain.

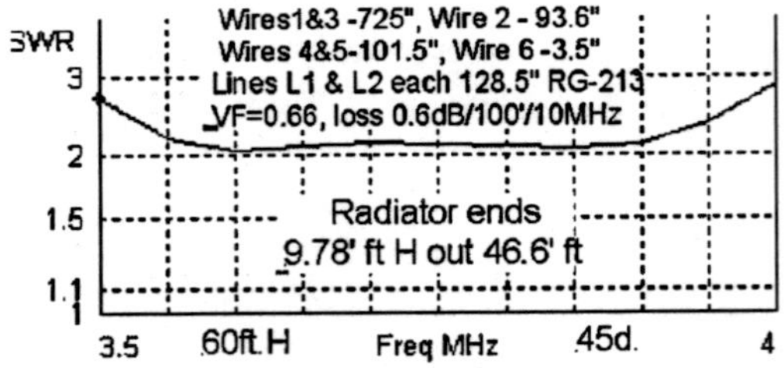

Figure 42

When this antenna is lowered down to 60' feet, as seen in Figure 42, surprisingly, there is very little change in the SWR across the entire bandwidth, even though this antenna still has a radiator slope of 45°. At a height of 60' feet the radiator ends are 9.78' feet above ground.

Looking at the Far-field gain and loss chart of Figure 43, you can see that reasonable RF gains across the entire 75-80m bandwidth are obtained. The most outstanding feature is that the RF losses are extremely low, almost to a point where there are none. This is attributable to having a large Delta, with wire lengths approaching 100" inches, having a very short capacitive Stub length and having these Stubs in series.

Far-field Gain & Losses at 89.6d.			
60ft.H	3.5	3.75	4.0MHz
No loss	3.76	3.99	3.89dBi
w/ loss	3.71	3.98	3.82dBi
RF Loss	0.05	0.01	0.07dB
Coax loss-RG-213 Vf=0.66, 60ft.H			

Figure 43

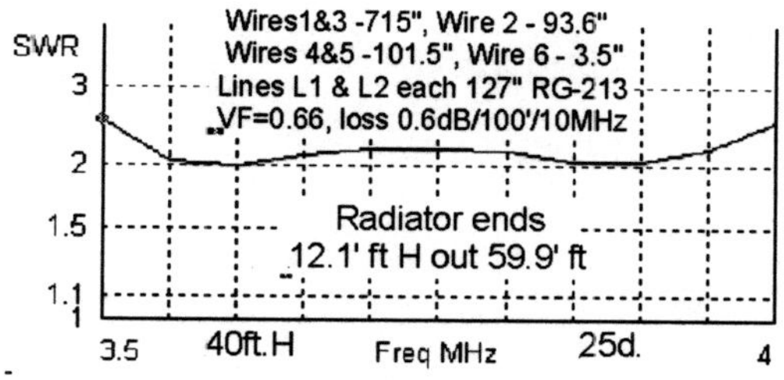

Figure 44

When the antenna is lowered to 45' feet as shown in Figure 44, it was necessary to change the radiator slope to 25° in order to keep the radiator ends well above ground at a height of 12.1' feet and extending outward to 59.9' feet. At this lower height, an SWR of approximately 2/1 can still be maintained across the entire 75-80m bandwidth.

Even though the height has been lowered, the Far-field RF gains, as shown in Figure 45 has changed very little. That is primarily due to reducing the antenna slope. The coaxial cable losses have also remained substantially the same, which are indeed very low.

Far-field Gain & Losses at 89.8d.			
40ft.H	3.5	3.75	4.0MHz
No loss	3.30	3.82	4.02dBi
w/ loss	3.24	3.81	3.96dBi
RF Loss	0.06	0.01	0.06dB
Coax loss-RG-213 Vf=0.66, 60ft.H			

Figure 45

This "Sd" Delta has all the qualities to be classified as a "Super-Delta" antenna and this designation which was given many years ago should indeed remain.

The previous book, titled "Super-Dipole Antennas" contains a large quantity of "Sd" variations and it is worthwhile to be aware of them. Up to this point, most of the configurations in the previous book and those in this Chapter have been dealing with configurations that required either 2 Capacitors or 2 Open-ended capacitive Stubs. In the next chapter all attempts will be made to utilize one Capacitor or Stub and many new configurations will be presented.

Super Max-Dipole Antennas

By Edward J. Shortridge W4JOQ

Chapter 2

Super-Diamond Inverted "V" Antennas

Introduction

Chapter 1 presented Inverted "V" antennas, which were Broad-Banded in a variety of configurations, all of which fit into the "Unified Broad-Banding Method".

The last antenna shown in Chapter 1 was a "Super-Delta", which has excellent characteristics. It did not use any discrete capacitors but instead, used two short coaxial capacitive Stubs in order to resonate the Delta near the center of the intended bandwidth. This antenna is quite simple and performs in an excellent manner.

Chapter 2 will continue further with configurations that allow greater versatility however these antennas will all use a discrete capacitor instead of "Capacitive Coaxial Stubs". This new configuration will be altered from a single Delta, into a dual Delta and will be called a "Diamond Antenna".

"S" Unified Broad-Banding System

This "Unified Broad-Banding Method" utilizes 2 resonant circuits in order to achieve Broad-Banding, one of which is the Radiator, while the other can be various forms of a parallel resonant circuit. These two resonant circuits must be isolated from each other and some means of controlled coupling must be used to provide the minimum of coupling that will produce "Critical Coupling", at which point a much wider SWR bandwidth can be achieved.

During many years of modeling "Inverted-V" Antennas, a suspicion has developed about adding a Delta or Diamond resonant circuit to an Inverted "V" Antenna. If the Delta or Diamond configurations are made so that they provide a Source of 12.5 or 25 Ohms instead of 50 Ohms, the configuration will be greatly reduced in physical size and better SWR bandwidths can be achieved.

You might rightfully ask, "Is this due to having to utilize a much larger value of capacitance in order to resonate this smaller Delta or Diamond configuration, or has the physical area that it covers somehow had an effect on the Coupling between the Radiator and the Broad-Banding circuit"? It is my belief that additional "Mutual Inductive Coupling" exists particularly in Inverted-V Antennas that are folded down at steep angles, such as 45°. When the Broad-Banding circuit is nestled up in the Inverted V, there is a very good chance of this additional coupling being present. Some of the antennas being presented will have the resonant capacitor end of the Diamond configuration, either at the bottom or the entire circuit rotated so that the capacitor is at the Apex. In each case where this was done, the resulting SWR plot showed that it had insufficient coupling to achieve the best SWR and therefore, the coupling had to be increased.

This should be a clear indication that moving the resonant Broad-Banding circuit outside (above) the confines of the Inverted-V reduced the mutual inductive coupling. Additionally, it appeared that the SWR was lower than previously.

After reading the remainder of this chapter, you will have a better understanding about this subject, so save your opinions until then.

Diamond 75-80m 25-50 Ohm 390 pF, "Series-section Matching"

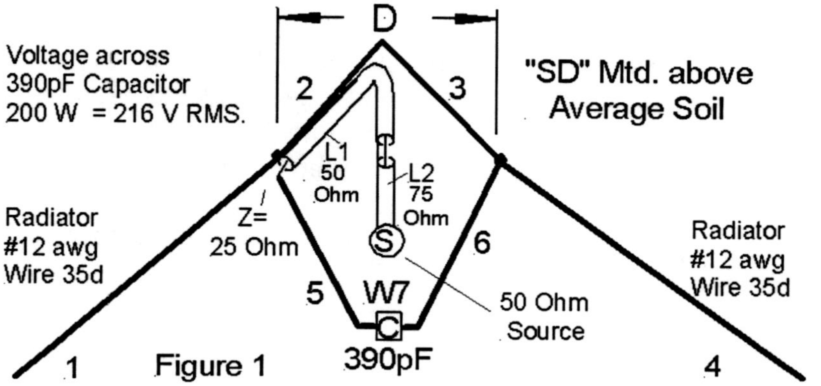

Figure 1

The antenna as shown in Figure 1 is in the shape of a Diamond, with a short Wire 7 in which the 390pF Load Capacitor has been added. It was designed to match 25 Ohms at the Feed-point located on the upper end of Wire 5. With a Radiator slope of 35° and mounted at a height of 85' feet above Average-soil, this antenna has a flat SWR of 2/1 across the entire 75-80m bandwidth, when measured at the 25 Ohm Source.

This Diamond configuration was a necessary result of choosing a low Source impedance of 25 or even 12.5 Ohms, along with this selection of a capacitance value of 390pF. While Wires 2, 3, 5 and 6 forms a loop that is resonated to the center of the desired bandwidth by the Fixed 390pF Capacitor, Wires 2 and 3 forms a common wire with the Radiator. Their total length is adjusted to achieve a common coupling, which is adequate to achieve "Critical-Coupling", which can produce the best SWR bandwidth.

If a larger capacitor value is selected, the length of Wires 5 and 6 will require shortening and if the value is large enough, Wires 5, 6 and 7 could get to a point where they are in a straight line, at which point the configuration will be a Delta and not a Diamond.

On the other hand, if the capacitor value is reduced substantially, Wire 7 and its connection to Wires 5 and 6, will be extended downward forming an elongated section with the angle between Wires 5 and 6 being an unusually narrow angle and in a shape that is quite unwieldy. With the current arrangement as shown in Figure 1, it has a shape that is more reasonably modeled and has better accuracy.

A 1/4wL 35 Ohm coaxial line can be used in order to match the 25 Ohms to 50 Ohms, but instead, a "Series-section Matching Line" was used to transform the impedance to a 50 Ohm Source. This line runs from the 25 Ohm Feed-point on the upper end of Wire 5, along Wire 2 to the Apex of the Inverted V, at which point it runs downward. In order to maintain a balanced feed, it is best if the braid of this line is connected to this Apex and then the line can continue downward. When using an RF transmitter power of 200 Watts the voltage across the capacitor is 216 Volts RMS. Therefore, there is little stress on the capacitor and higher operating power levels can be used without voltage breakdown.

Six Diamond Antennas of this type were modeled using a variety of capacitor values from 180 to 560pF and most of them were able to achieve approximately 2/1 SWR across the bandwidth, such as was achieved when using a 25 Ohm source at the upper end of Wire 5. When either the 35 Ohm 1/4wL Coaxial matching line or a "Series-section Coaxial Matching Line" is used, this particular antenna above all other Diamond Antennas has an unusual characteristic.

When either of these two types of matching lines is used with this antenna, they seem to create "Double Broad-Banding" producing a much better SWR bandwidth, which none of the other Diamond configurations can achieve. <u>This is phenomenal and occurs on very rare occasions.</u>

In order to preserve the physical shape of a Diamond configuration, a dimension "D" which is shown on Figure 1 must be maintained. An insulated wire or a small diameter fiberglass rod should be used. As the radiators are pulled down at the proper sloping angle, there is a tendency for this dimension to be pulled apart and this method of maintaining the dimension seems quite necessary.

When this antenna has a Source at the 25 Ohm Feed -point (Z), it has a very flat SWR response of 2/1 across the entire 75-80m bandwidth, as shown in Figure 2. This is an SWR that is quite typical of most of the Diamond Antennas; however this SWR can be improved.

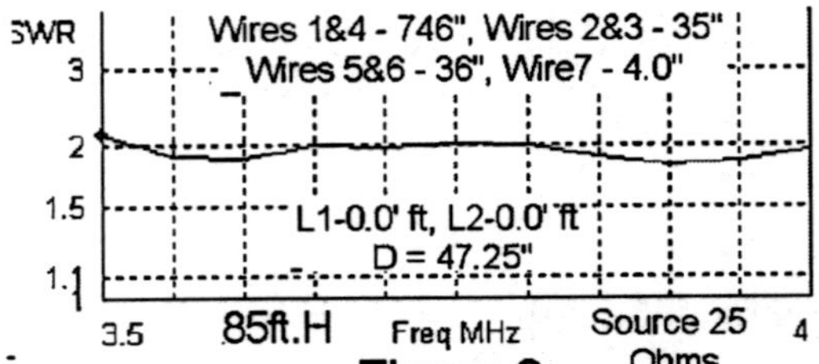

Figure 2

After adding the "Series-section Matching Line" from 25 to 50 Ohms, a greatly improved SWR bandwidth which was previously in the order of an SWR of 2/1 is now improved to approaching an SWR of 1.5/1 across the center of the bandwidth, as shown in Figure 3. This SWR plot appears to be very difficult to duplicate on other antenna configurations when they are mounted at the same height and have a Radiator slope of 35°.

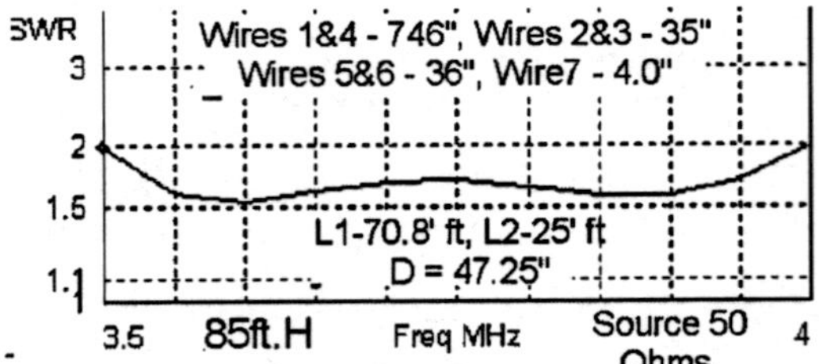

Figure 3

Far-field Gain & Losses at 55.1d.			
85ft.H	3.5	3.75	4.0MHz
no loss	5.19	5.34	5.33dBi
w/ loss	4.73	5.01	4.75dBi
RF Loss	0.46	0.33	0.586dB
Coax-0.5dB/100'/10MHzVf=0.84			

Figure 4

The Far-field RF radiation gain and losses chart of Figure 4 shows that this antenna has good gains of over 5 dBi and at the center of the 75-80m bandwidth, it has the maximum radiation angle of 55.1°. The RF losses in the 50/75 Ohm Coaxial cables are higher than necessary. The "Series-section Matching" is not well-suited for matching a 25 Ohm Feed-point impedance to 50 Ohms.

This requires a total Coaxial cable length of 95' feet, whereas other matching configurations such as these use a 1/4wL. A 35 Ohm matching line will be approximately 1/2 of this length and thereby reducing the overall coaxial cable losses.

When this antenna, is lowered to 65' feet, the SWR begins to suffer, but it still averages less than 2/1 SWR across the entire 75-80m bandwidth as shown in Figure 5. Each time the antenna is lowered in height, it is necessary to shorten the Radiator ends due to the increased capacitance to ground.

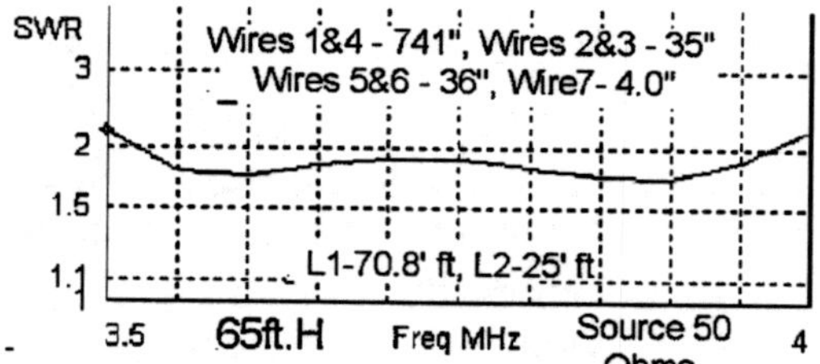

Figure 5

An important consideration: The 390pF capacitor should be chosen with at least 5% tolerance and preferably better. The modeling of this antenna was accomplished using precisely 390 pF and as a practical consideration, the exact capacitor will deviate slightly from the 390 pF and the SWR will be slightly uneven or more so depending on the tolerance of the capacitor.

In order to overcome this difficulty Wire 7 (W7), while attached to Wires 5 and 6 should be moved upward or downward in order to change the length of Wires 5 and 6, to once again retain the SWR symmetry.

Wire 2, which along with all of the other wires in this configuration are specified as #14 AWG wire, however after adding the L1 coaxial cable alongside of it and having an attached braid to Wire 2, can result in lower coupling between the Radiator and the resonant Broad-Banding circuit. Since it is a larger diameter and appears to have less inductance than the #14 Awg Wire 2. Consequently, Wires 2 and 3 should be increased in length slightly.

In order to do this, the upper ends of Wires 2 and 3 should be raised in 1" inch increments, so that the original coupling is restored, as indicated by the shape of the SWR plot.

Continuing further beyond Figure 5, the Far-field radiation gain and losses chart of Figure 6 shows a modest reduction in radiation gain due to the change in height from 85' to 65' feet. There is also a slight rise in 89.6° coaxial cable losses.

Far-field Gain & Losses at 89.6°			
65ft.H	3.5	3.75	4.0MHz
No loss	5.02	5.28	5.22dBi
w/ loss	4.49	4.93	4.54dBi
RF Loss	0.51	0.35	0.68dB
Coax-0.5dB/100'/10MHzVf=0.84			

Figure 6

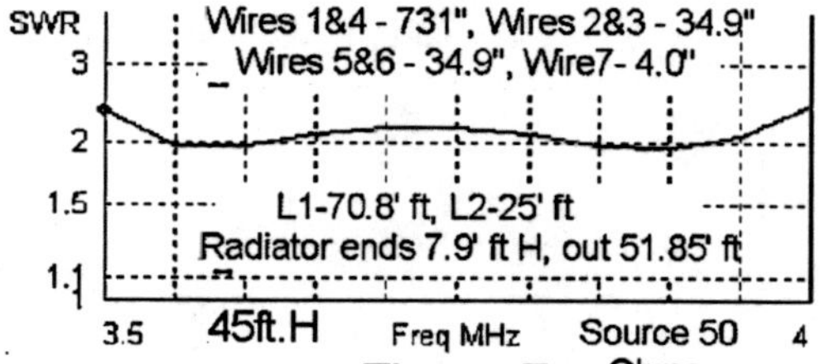

Figure 7

The antenna was mounted even lower at a height of 45' feet, as seen in Figure 7 and the SWR bandwidth deteriorated very slightly, too approximately 2/1 across the entire bandwidth and this is better than what could be achieved by other configurations.

The Far-field radiation gain chart of Figure 8 shows a distinct drop in RF gain, which should be expected. The angle of maximum gain is still at approximately 90° and the coaxial cable losses have not changed.

Far-field Gain & Losses at 89.8°			
45ft.H	3.5	3.75	4.0MHz
No loss	3.16	3.60	3.74dBi
w/ loss	2.65	3.24	3.06dBi
RF Loss	0.51	0.36	0.686dB
Coax-0.5dB/100'/10MHzVf=0.84			

Figure 8

This antenna has extremely good characteristics and the RF gains might be reduced due to the "Series-section Matching" cable losses but other means can be used to match the 25 Ohm Feed-point impedance to 50 Ohms and these other methods will have considerably less RF radiation loss.

Diamond 75-80m 25-50 Ohm 390 pF, 1/4wL Open-end Line

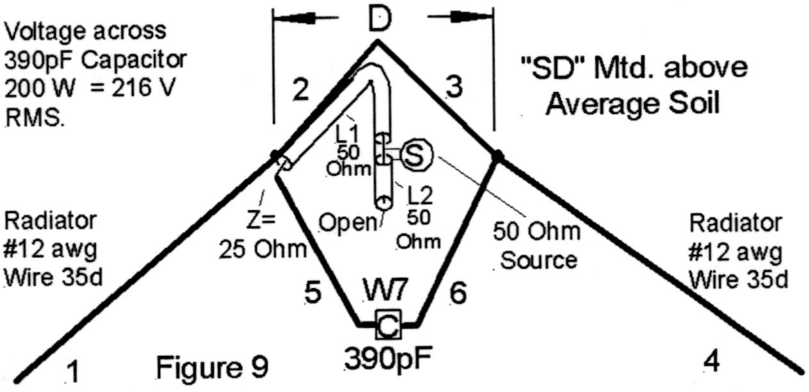

Figure 9

The previous Diamond Antenna which used "Series-section Matching" was used for this antenna but the 25 to 50 Ohm matching section was changed from a Series-section configuration to a 1/4wL Open-ended 50 Ohm coaxial line, which has a 50 Ohm Source tap down somewhat from the center, as shown in Figure 9. In doing this, the length of the coaxial line which was in the order of 95' feet in length, was reduced to a total length of 42' feet. Line L1 is now 24' feet in length, while L2 is 18' feet. The previous Series-section line was much longer than necessary and it has now been reduced in length to less than one-half. This will reduce the coaxial losses to a very large extent, even though an edited transmission line will be required. This is necessary since the Source is tapped-up on the line, instead of being connected to the very end of this line.

With this new matching line in place and the antenna mounted at a height of 85' feet, the SWR plot of Figure 10 shows a good SWR that is well below 2/1, however this SWR has deteriorated slightly from the previous antenna and most of this may be due to having less coaxial loss, since lines L1 and L2 have been greatly reduced in length.

The Far-field RF gains are typical for this antenna height, as seen in Figure 11, averaging approximately 5.2dBi, while the RF cable losses have been reduced to 0.19, 0.10 and 0.27db across the entire bandwidth.

Far-field Gain & Losses at 55d.			
85ft.H	3.5	3.75	4.0MHz
no loss	5.22	5.21	5.35dBi
w/ loss	5.03	5.11	5.08dBi
RF Loss	0.19	0.10	0.276dB
Coax-0.5dB/100'/10MHzVf=0.84			

Figure 11

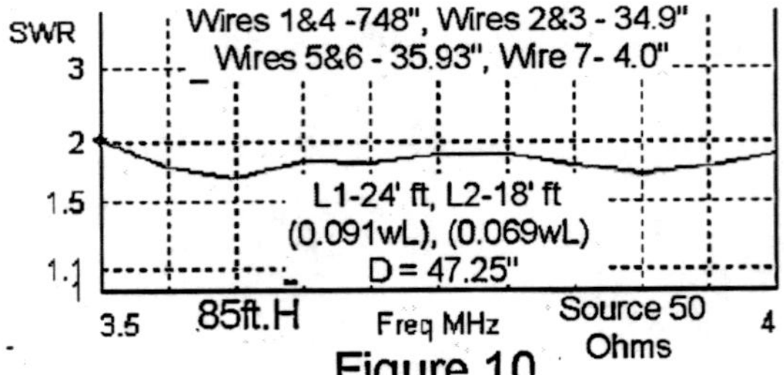

Figure 10

Diamond 75-80m 25-50 Ohm 390 pF, Rotated, Open Line

The previous Diamond Antenna was configured with the 390pF Capacitor at the bottom end of the Broad-Banding circuit and with the matching section coaxial cable running from the upper end of Wire 5 up to the Apex of Wires 2 and 3 and then proceeding downward. As the coaxial cable went downward, Wire 7 (W7) with the 390pF Capacitor attached had to be secured either to the downward coaxial cable or to the Mast, in order to maintain its position. This was somewhat of an awkward arrangement which needed some improvement.

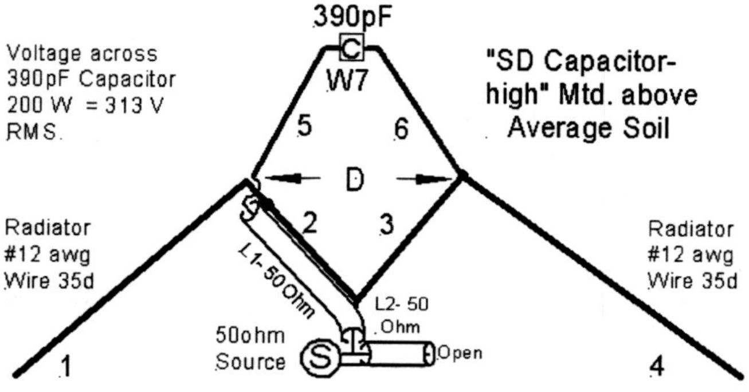

Figure 12

By rotating the Diamond, which consists of Wires 2, 3, 5, 6 and 7 180° on the pivot points of radiator Wires 1 and 4, the 390pF Capacitor is now located at the top of the configuration. One of the coaxial feeds runs down along Wire 2 to the bottom end of Wires 2 and 3 at which point it should extend down quite a long distance to the Source and line L2 which is much further down to an Open-end of that line. This arrangement can be seen in Figure 12.

With this arrangement, the weight of the coaxial cables will maintain the Diamond shape, as long as there is an insulated cord or fiberglass rod used to maintain the dimension "D".

The top of the radiator still consists of Wires 2 and 3 but these have been lowered slightly too where they are at the lower portion of the Diamond configuration. This will result in a slightly lower radiator and it is expected that the Far-field RF gain will change slightly.

When the SWR was plotted, it was immediately noted that the SWR curvature indicated that there was under-coupling between the radiator and the Diamond Broad-Banding resonant circuit. Therefore, it was necessary to increase each of the Wire lengths of 2 and 3 from 34.9" inches to 38.13" inches and then Wires 5 and 6 had to be altered from 35.9" up to 37.5" inches in length.

All of the Diamond wire lengths were increased, while still using a capacitor value of 390pF. After making these changes the SWR plot of Figure 13 shows a notable improvement in the SWR which is approaching 1.5/1 across the major portion of the bandwidth, making this antenna unusually improved, above the standard SWR of 2/1 of many of the other configurations.

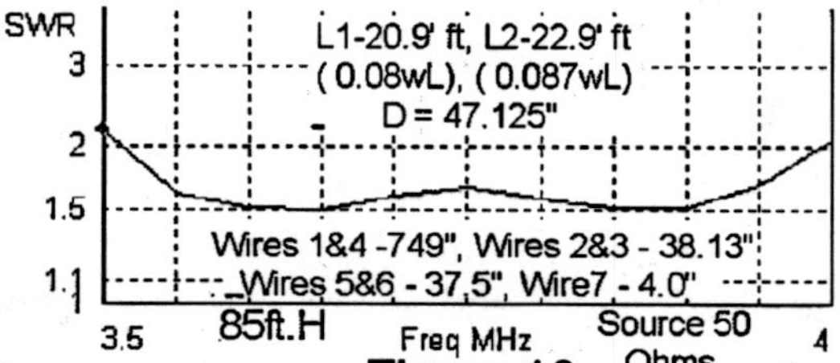

Figure 13

The RF voltage across the 390pF Capacitor is always a big concern but in this antenna if an RF power level of 200 Watts was used, the resulting voltage across this Capacitor will be 202V RMS. This is extremely low and if the Capacitor voltage rating is 1000 volts, then much higher RF power can be used.

The Far-field RF gain and losses chart of Figure 14 shows that this antenna has very good RF gain, even though Wires 2 and 3 of the radiator are lower than the previous antenna. Although the 1/4wL Open-ended coaxial cable has been greatly reduced in length, compared with the "Series-section Matching Line", there is still approximately 44' feet of coaxial cable used in this line. Nevertheless, the RF losses are quite low and this antenna could be considered a very efficient antenna.

Far-field Gain & Losses at 55d.			
85ft.H	3.5	3.75	4.0MHz
no loss	5.29	5.35	5.37dBi
w/ loss	5.03	5.25	5.07dBi
RF Loss	0.26	0.10	0.30dB
Coax-0.5dB/100'/10MHzVf=0.84			

Figure 14

This antenna performs extremely well; providing superb SWR values and is well suited to being an "Inverted V". It's Far-field RF radiation gain is slightly better than most and its coaxial cable losses are low, making this an ideal antenna which should be considered a "Super-Dipole Antenna".

Diamond 75-80m 25-50 Ohm Rotated, "Series-section Matching"

Although it has been demonstrated that the length of a "Series-section Matching Line" might be extremely long when matching a 25 Ohm Feed-point impedance to a 50 Ohm Source, we are once again using it however in this case the Diamond resonant circuit has been rotated with the 390pF Capacitor located at the top end of the Apex. The prime purpose is to show how this combination will affect the SWR across the entire bandwidth.

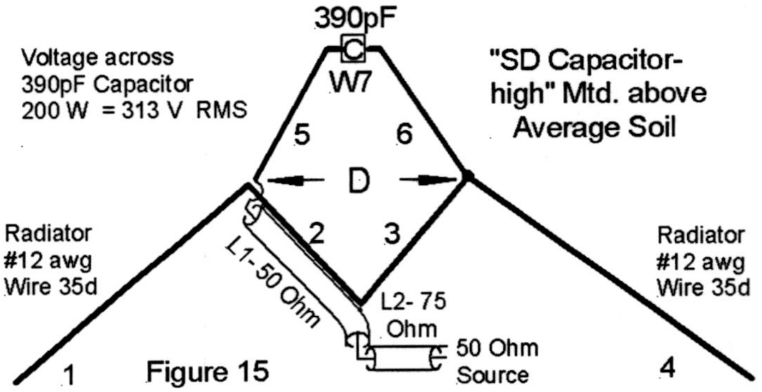

Figure 15

The configuration of this antenna, as shown in Figure 15, has the same shape as the previous antenna but some of the wire lengths may differ slightly. The "Series-section Matching Line" is very similar to the previous antenna and it uses a 59' foot length of 50 Ohm coaxial cable for line L1 and a 42' foot length of 75 Ohm cable for line L2, as shown in Figure 16.

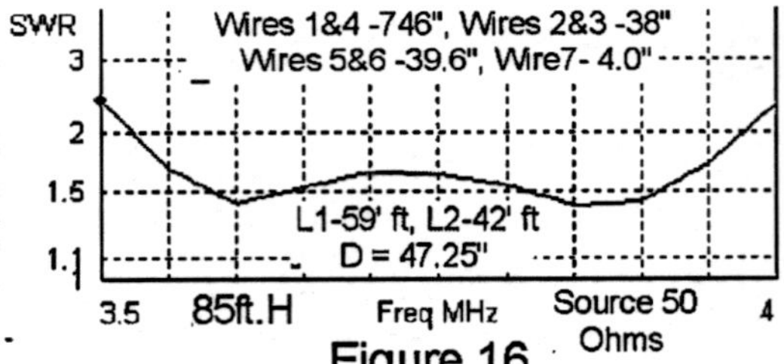

Figure 16

This has resulted in an extremely good SWR averaging close to 1.5/1 over a good portion of the bandwidth and deteriorates to slightly above 2/1 at the band edges. This confirms that if a "Series-section" line is used with a rotated configuration, very good SWR outcomes can still be realized when the antenna is mounted at a height of 85' feet.

The Far-field radiation chart of Figure 17 indicates that there is slightly less RF gain compared to the previous antenna that uses the 1/4wL open-ended matching line. Also as would be expected the longer "Series-section Matching Line" has introduced more coaxial cable losses.

Far-field Gain & Losses at 55.4d.			
85ft.H	3.5	3.75	4.0MHz
no loss	5.37	5.50	5.52dBi
w/ loss	4.89	5.14	4.79dBi
RF Loss	0.48	0.36	0.736dB
Coax-0.5dB/100'/10MHzVf=0.84			

Figure 17

Lowering the antenna to 65' feet has caused the SWR to deteriorate slightly, as shown in Figure 18 but it is still excellent and far better than most antennas mounted at a similar height and using a Radiator slope of 35°.

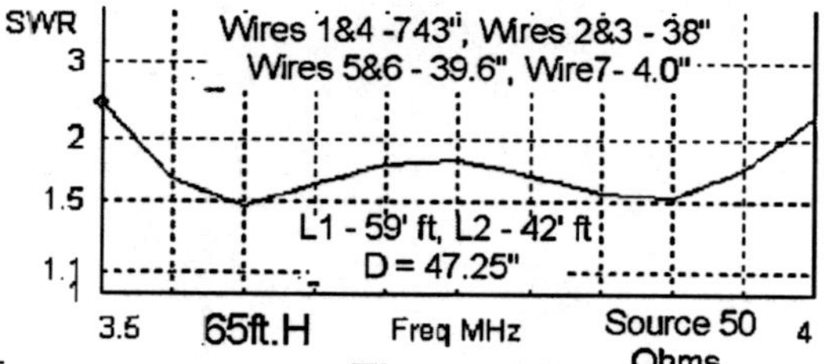

Figure 18

The Far-field RF gains as shown in Figure 19 have been reduced slightly but are almost the same as those at 85' feet. On the other hand, the RF coaxial losses appear to have increased slightly.

Far-field Gain & Losses at 89.6d.			
65ft.H	3.5	3.75	4.0MHz
no loss	5.36	5.57	5.39dBi
w/ loss	4.70	5.21	4.59dBi
RF Loss	0.66	0.36	0.80dB
Coax-0.5dB/100'/10MHzVf=0.84			

Figure 19

Lowering the antenna further to a height of 45' feet has caused further deterioration of the SWR, as shown in Figure 20 and it averages with an SWR of 2/1, which is a value that similar antennas may only achieve at a mounting height of 85' feet or higher.

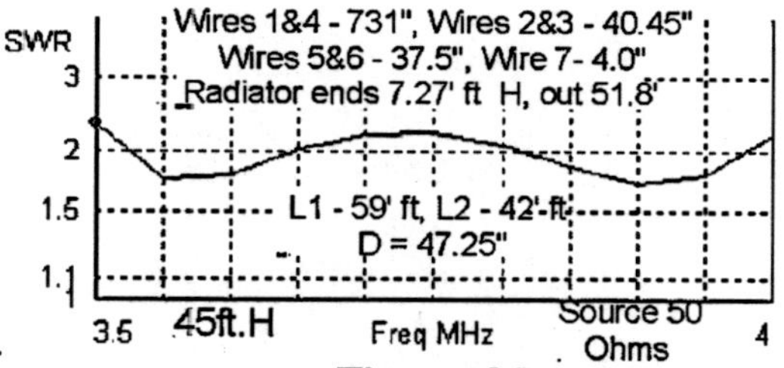

Figure 20

The SWR as shown in Figure 20 has increased, but due to the coaxial losses it is slightly lower than normal and these coaxial losses appear to have increased very slightly as shown in Figure 21.

Far-field Gain & Losses at 89.7d.			
45ft.H	3.5	3.75	4.0MHz
No loss	3.17	3.65	3.79dBi
w/ loss	2.50	3.28	2.98dBi
RF Loss	0.67	0.37	0.816dB
Coax-0.5dB/100'/10MHzVf=0.84			

Figure 21

Although this antenna has very good SWR bandwidths, its overall radiation has dropped due to the very long "Series-section Matching Line", which is not well adapted to matching 25 Ohms impedance to a Source impedance of 50 Ohms without having excessive length. There are other matching configurations that can provide lower losses and thus higher RF gains. As a result, this antenna is not recommended as a "Super-Dipole".

Diamond 75-80m 25-50 Ohm Rotated, 1/4wL 35 Ohm line

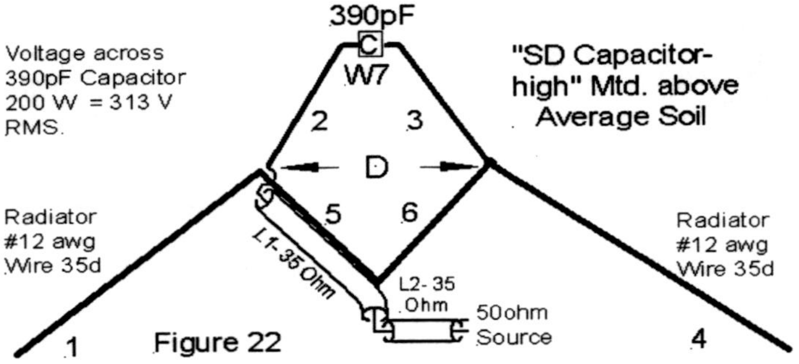

Figure 22

Continuing with the use of the Diamond rotated antenna that has a 25 Ohm Feed-point, a 1/4wL 35 Ohm matching line is used instead of the "Series-section Matching Line". When this line is used, there is no need to tap the 50 Ohm Source onto the line but instead it is connected to the bottom end of this line as shown in Figure 22. The total length of this line has now been reduced to 43.27' feet in length and the coaxial losses will be decreased, thereby providing slightly increased Far-field RF gains.

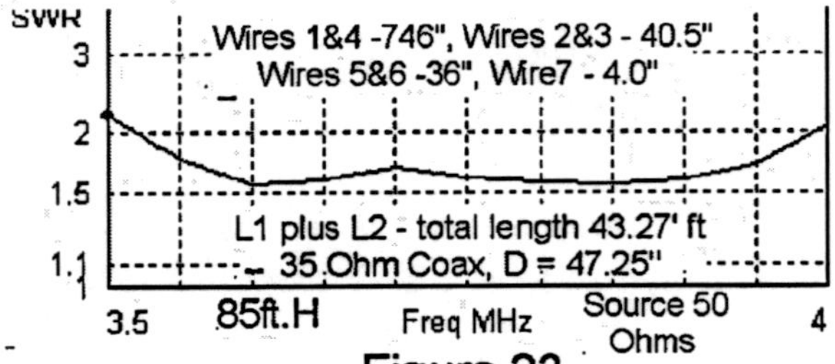

Figure 23

The SWR plot of Figure 23 clearly shows that the low SWR is still maintained whenever the 390 pF Capacitor is still used. The SWR is slightly above 1.5/1, for much of the bandwidth and the band edges are at approximately 2/1.

The wire lengths of the Diamond resonant circuit have been changed very slightly in order to achieve the best SWR. Once again, it must be emphasized that the dimension "D" must be secured in order to maintain a good SWR response.

The chart of Figure 24 clearly shows the expected improvements of using a shorter, less lossy matching line. The Far-field RF radiation gains are over 5.0dBi across the entire bandwidth and the RF coaxial losses have been reduced to 0.24, 0.18 and 0.27dB across the entire 75-80m bandwidth. This has resulted in having the entire 35 Ohm matching line of 43.27' feet also act as part of the transmission line length. Everything has culminated in an antenna which can clearly be classified as a "Super-Dipole Antenna".

Far-field Gain & Losses at 89.7d.			
85ft.H	3.5	3.75	4.0MHz
no loss	5.24	5.27	5.39dBi
w/ loss	5.00	5.09	5.12dBi
RF Loss	0.24	0.18	0.27dB
Coax-0.6dB/100'/10MHzVf=0.66			

Figure 24

As this antenna is reduced in height to 65' feet, the SWR has deteriorated slightly, but is still below an SWR of 2/1, as shown in Figure 25. Unfortunately, most antennas will have a deteriorated SWR as their heights are lowered.

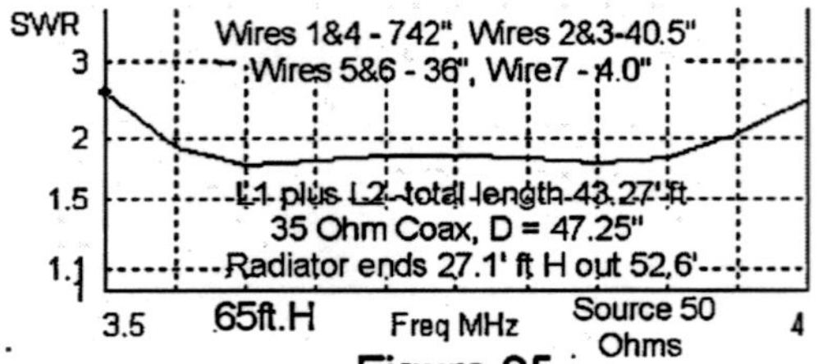

Figure 25

The Far-field radiation chart of Figure 26 shows a change in the RF gain, with an accompanying increase in the coaxial matching losses. There appears to be no means of avoiding these changes. At a mounting height of 45' feet, as seen in Figure 27, the 35° slope of the radiators will bring the ends of the radiators down to a height of 7.5' feet above ground and they will be out 51.93' feet from the antenna center. The SWR bandwidth has deteriorated to very slightly above an SWR of 2/1. It is good to remember that the typical Diamond Antenna that uses other than 390pF as a Capacitor, normally will have difficulty achieving this SWR when mounted at a height of 85' feet. No one likes deterioration but in reference to the other antennas this antenna is very good.

Far-field Gain & Losses at 89.7d.			
65ft.H	3.5	3.75	4.0MHz
no loss	5.08	5.34	5.27dBi
w/ loss	4.82	5.14	4.98dBi
RF Loss	0.26	0.20	0.29dB
Coax-0.6dB/100'/10MHzVf=0.66			

Figure 26

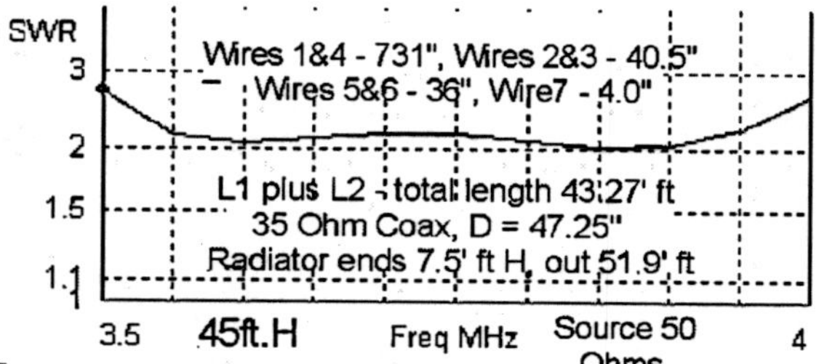

Figure 27

The RF gain and losses chart of Figure 28 shows RF gains that are well above average for antennas at this height and the RF coaxial losses have increased slightly.

This "Diamond Super-Dipole Antenna" appears to be a winner and considering its physical size and lack of complexity, it is an antenna to be proud of and it would be very difficult to find an antenna with qualities equal to it.

Far-field Gain & Losses at 89.7d.			
45ft.H	3.5	3.75	4.0MHz
no loss	3.17	3.65	3.79dBi
w/ loss	3.10	3.59	3.70dBi
RF Loss	0.07	0.06	0.09dB
Coax-0.6dB/100'/10MHzVf=0.66			

Figure 28

Diamond 75-80m 12.5-50 Ohm, Rotated 1/4wL 25 Ohm line

All of the previous antennas in Chapter 2 were configured to have a Feed-point impedance of 25 Ohms. In this antenna the rotated Diamond configuration is still used, but it is modified to have a Feed-point impedance of 12.5 instead of 25 Ohms. In order to accomplish this, the load Capacitor of the 390pF was changed to 560pF. In order to have the Diamond which consists of Wires 2, 3, 5 and 6 to once again resume in the center of the 75-80m band, all of these wire lengths had to be greatly reduced in length.

With this new configuration, the Feed-point was moved from Wire 5 to the lower end of Wires 2 and 3. At this point the desired impedance will be 12.5 Ohms. In order to match this impedance to 50 Ohms, a 1/4wL, 25 Ohm matching-line can be used. This impedance line is not available; therefore two 50 Ohm coaxial cables can be combined in a parallel configuration. A 50 Ohm coaxial cable with a specified loss of 1.0dB/100'/10MHz can be used and these two combined cables will have a total loss of 0.5dB, instead of the 1.0dB loss.

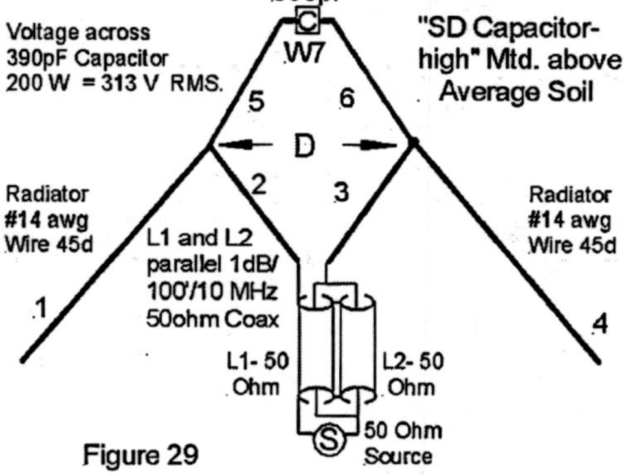

Figure 29

This new configuration, as shown in Figure 29, has its radiator ends sloping at 45° instead of 35° which was used in the previous antenna. This change was made in anticipation that this new configuration will provide a satisfactory SWR bandwidth that will make it a practical antenna.

When mounted at a height of 85' feet and using 43.27' feet of coaxial cable for each of the Lines L1 and L2, which are in parallel. An SWR averaging somewhat less than 2/1 was obtained across the entire bandwidth, as shown in Figure 30. Considering the steep angle of the radiators, at a slope of 45°, this is a very good SWR. Remember each of the two parallel coaxial cables has a specified loss of 1.0dB/100'ft/10MHz equaling a loss of 0.5dB.

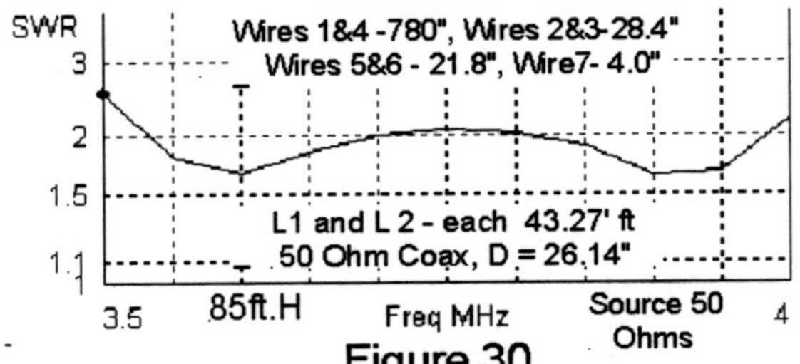

Figure 30

The Far-field RF radiation gain and losses chart of Figure 31 shows gains that are quite reasonable when using a 45° radiator slope. The coaxial losses of 43.27' feet of matching line, which also acts as part of the transmission line, are 0.47, 0.15 and 0.42dB across the 75-80m band. Overall these losses are very modest and this antenna should perform well.

Far-field Gain & Losses at 58.3d.			
85ft.H	3.5	3.75	4.0MHz
no loss	4.71	4.73	4.40dBi
w/ loss	4.24	4.58	3.98dBi
RF Loss	0.47	0.15	0.42dB
Coax-1.0dB/100'/10MHzVf=0.84			

Figure 31

Lowering the antenna to 65' feet caused only minor degradation of the SWR bandwidth and it still averages slightly under an SWR of 2/1, as shown in Figure 32. Each time the antenna is lowered, the radiator Wires 1 and 4 are reduced in length, in order to re-center the SWR bandwidth. With this change in height, the Far-field RF gain has dropped a modest amount, while the coaxial cable losses have increased very slightly, as can be seen in Figure 33.

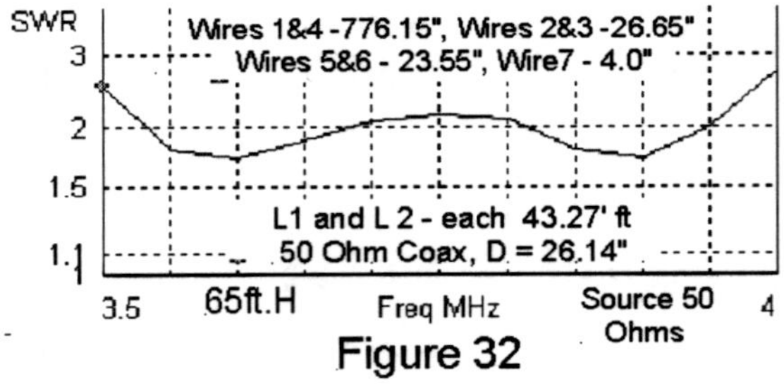

Figure 32

Far-field Gain & Losses at 89.9d.			
65ft.H	3.5	3.75	4.0MHz
no loss	4.47	4.48	4.19dBi
w/ loss	4.01	4.31	3.70dBi
RF Loss	0.46	0.17	0.49dB
Coax-1.0dB/100'/10MHzVf=0.84			

Figure 33

Lowering the antenna further to a height of 54.5' feet, the SWR has deteriorated and is now averaging slightly above at 2/1 and the band edges have deteriorated to an SWR of 3/1 as can be seen in Figure 34.

When using a radiator slope of 45°, this antenna can only be lowered to 54.5' feet and this will place the ends of the radiators at 7.5' feet above ground and out at a distance of 46.18' feet. With an antenna slope of 35°, the antenna can be lowered to 45' feet and still have the radiator ends safely out of reach.

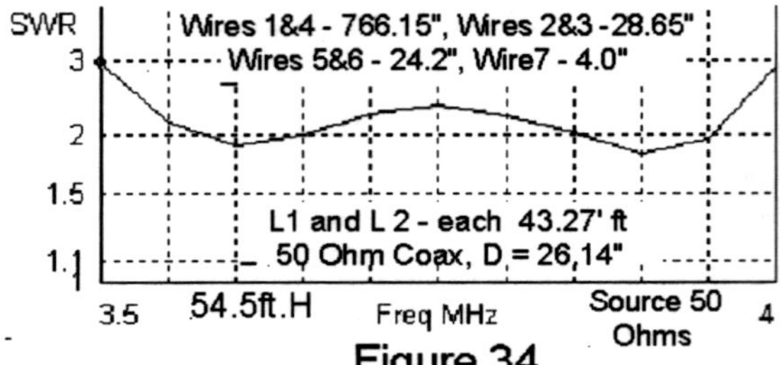

Figure 34

Far-field Gain & Losses at 89.9d.			
54.5ft.H	3.5	3.75	4.0MHz
no loss	3.43	3.77	3.43dBi
w/ loss	2.95	3.57	2.97dBi
RF Loss	0.48	0.20	0.46dB
Coax-1.0dB/100'/10MHzVf=0.84			

Figure 35

The Far-field RF radiation chart of Figure 35 indicates that this antenna at the height of 54.5' feet still has good RF gains, in the order of 3dBi and the coaxial losses have changed very little, making this a very good antenna that should work well.

Even though the SWR was restricted due to the 45° radiator angle slope, all of its performance is quite good but it could be greatly improved and be able to be mounted down to 45' feet above ground if a 35° radiator slope is used.

Diamond 75-80m 12.5-50 Ohm, 12.5/50 Balun - 45° slope

The previous antenna utilized a 25 Ohm 1/4wL matching line in order to match the 12.5 Ohm feed impedance located at the bottom of Wires 2 and 3. However, this antenna will use the same matching technique; however it will utilize a 12.5/50 Ohm Step-up Balun instead, as shown in Figure 36.

In order to have a 12.5/50 Ohm Balun, the internal lines should ideally be 25 Ohms. This cannot be achieved by utilizing close-spaced #14 AWG wire and the use of a small diameter 25 Ohm coaxial cable is not very practical. Therefore, other means must be utilized.

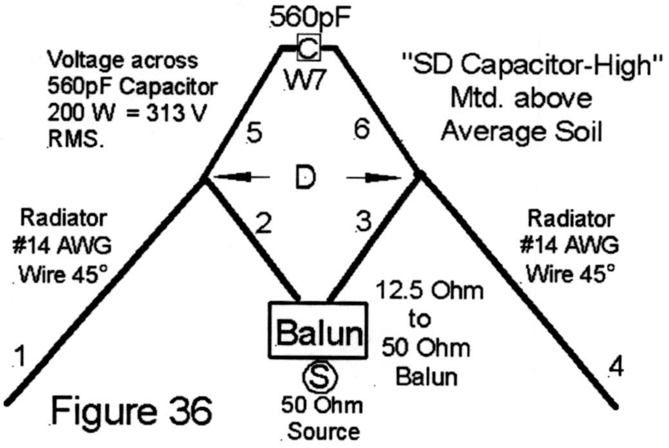

Figure 36

Many of the 50/200 Ohm Baluns in use today utilizes approximately 22" inches of Twin #14 AWG wire in each of the two windings. When these two wires are spaced relatively close to each other they have an impedance of approximately 100 Ohms, which is the desired internal line impedance for a 50/200 Ohm Balun.

An internal line length of 22" inches is a very short portion of a 1/4wL resonant line in the 75-80 or 160m bands, as used in the typical 50/200 Ohm Balun and if this typical Balun is used to transform a 12.5 Ohm impedance to 50 Ohms, there will be very minor undesirable effects that can be readily overcome by making very small adjustments to the antenna wire lengths.

A typical Balun is a four port device and the high impedance end does not necessarily need to be considered the balanced end but instead, the low impedance end can be connected to the 12.5 Ohm antenna impedance and is every bit as balanced as the high impedance end might be. A coaxial cable can be connected to the high impedance end of the Balun without any detrimental effects, even though the coaxial cable braid is normally connected to ground.

Such a Balun was used, as shown in Figure 36 and the SWR plot of Figure 37 shows that a very good SWR of approximately 2/1 was obtained across the entire bandwidth. This was obtained with the typical Balun but used in a reverse direction with the low impedance at the top end and the high impedance of 50 Ohms at the lower end.

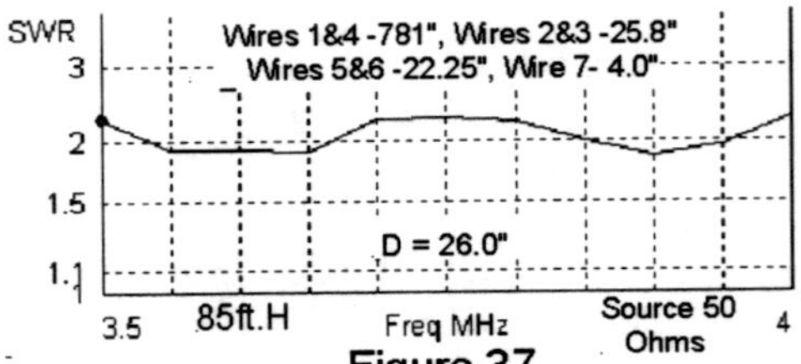

Figure 37

This SWR is not quite as good as the previous antenna, as shown previously in Figures 29 and 30. That antenna included a 25 Ohm 1/4wL matching line of slightly over 43' feet in length, whereas this antenna, as modeled, did not include any matching line but only a loss of normal 1.0dB/100'ft/10MHz for the 22" inches of 100 Ohm internal Balun lines.

Far-field RF Gains at 57.8d.			
85ft.H	3.5	3.75	4.0MHz
no loss	4.68	4.93	4.63dBi

Figure 38

The Far-field RF gain chart of Figure 38 shows gains of 4.68, 4.93 and 4.63dBi across the 75-80m bandwidth and these gains are quite good considering that this antenna has a radiator slope of 45° which tends to lower the expected RF gains.

The Apex of this antenna was lowered to 54.5' feet above ground, at which point the radiator ends were at 7.6' feet above ground, well out of reach and extending out to 46.3' feet. This is the lowest height that a 45° radiator slope can be used in the 75-80m band.

At this height the SWR obtained an average somewhat less than 2.5/1, as can be seen in Figure 39. Considering that the antenna has been lowered a total of 31.5' feet, the SWR has not suffered very much.

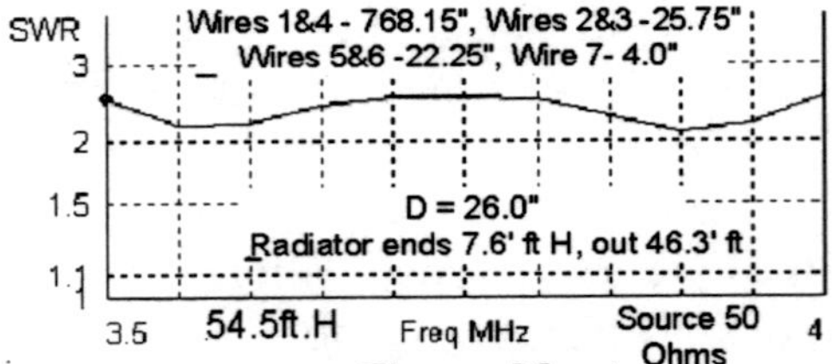

Figure 39

The Far-field RF gains, as shown in Figure 40 are quite good at 3.31, 3.54 and 3.33dBi across the 75-80m band, making this a very workable antenna that will perform well.

Far-field RF Gains at 89.8d.			
54.5ft.H	3.5	3.75	4.0MHz
no loss	3.31	3.54	3.33dBi

Figure 40

This antenna, which uses a radiator slope of 45°, has the constricted SWR bandwidth and restricted RF gains and therefore, the following antenna will be the same basic antenna except that the radiator slope will be adjusted for 35°, which will show an overall improvement.

Diamond 75-80m 12.5-50 Ohm, 12.5/50 Balun - 35° slope

The previous antenna which had a radiator slope of 45° was modified to reduce the slope to 35° and this configuration is shown in Figure 41. The radiator Wires 1 and 4 had to be decreased in length from 781" inches down to 773.15" inches in order to obtain a symmetrical SWR plot.

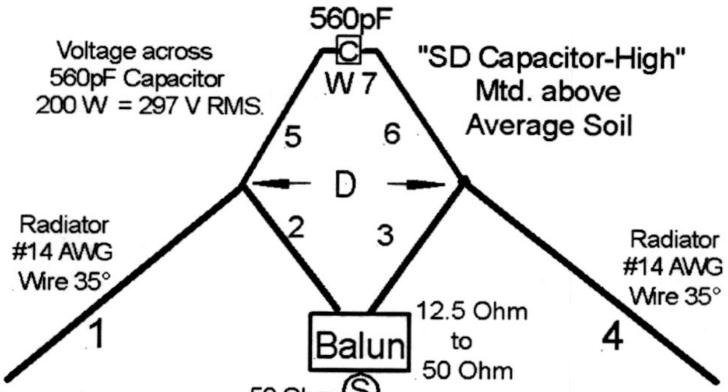

Figure 41

When mounted at an 85' foot height the SWR plot of Figure 42 shows a definite improvement in the SWR compared to the previous antenna that has a steeper 45° slope. The previous SWR of Figure 37 averages slightly above an SWR of 2/1, whereas the 35°slope radiator antenna has an SWR of 1.5/1 across the center of the bandwidth and deteriorating slightly above 2/1 at the band edges, which is a definite improvement.

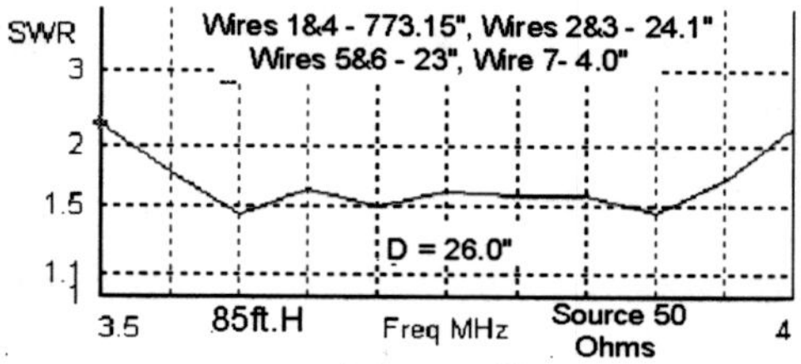

Figure 42

The Far-field RF gain chart of Figure 43 shows very good SWR gains of 5.11, 5.18 and 4.95dBi across the 75-80m band and it has a low radiation angle of 55.2° on the frequency of 3.75MHz.

Far-field RF Gains at 55.2°			
85ft.H	3.5	3.75	4.0MHz
no loss	5.11	5.18	4.95dBi

Figure 43

Lowering the antenna to a 65' foot height, required shortening the radiator lengths from 773.15" inches to 767" inches, to once again center the SWR bandwidth which has deteriorated somewhat but averages less than 2/1, as can be seen in Figure 44.

The Far-field RF gains, as shown in Figure 45, have deteriorated very slightly, but the angle of radiation at 3.75MHz has changed from 55.2° to 89.6°. This is a typical radiation angle however the RF gains are better than many of the other antennas when mounted at this height.

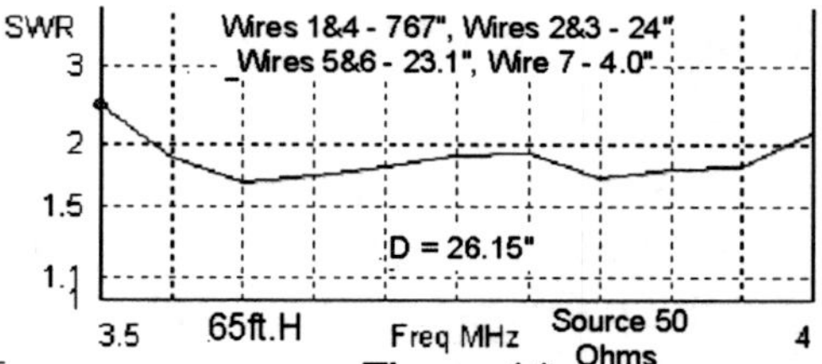

Figure 44

Because this antenna has a radiator slope of 35°, it is possible to lower this antenna to 45' feet above ground, whereas the antennas with a normal 45° slope are limited to a height of 54.5' feet above ground. At this height the radiator ends are 7.0' feet above ground and extend out 52.8' feet from the antenna center.

Far-field RF Gains at 89.6°			
65ft.H	3.5	3.75	4.0MHz
no loss	5.03	5.22	4.79dBi

Figure 45

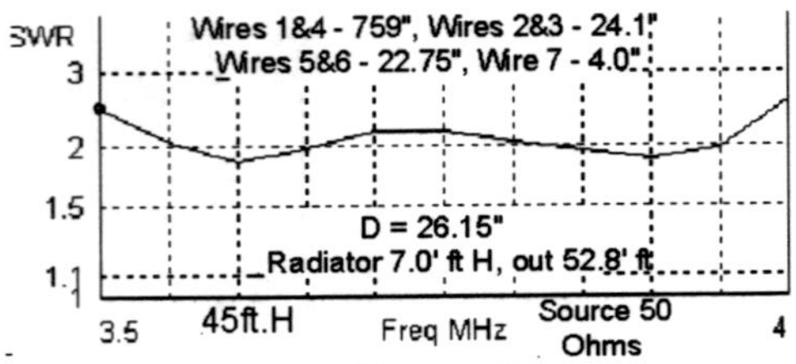

Figure 46

At a height of 45' feet, the SWR of Figure 46 indicates that the SWR averages 2/1 across the entire bandwidth, whereas the previous antenna which utilized a 45° slope and mounted higher at 54.5' feet had a noticeably deteriorated SWR well above 2/1. The lower angle of 35° has made this difference.

The RF gain chart of Figure 47 shows the Far-field RF gains averaging slightly above 3dBi across the entire 75-80m bandwidth. These gains have held up very well, making this a very good antenna, fully capable of performing in an excellent manner.

Far-field RF Gains at 89.7°			
45ft.H	3.5	3.75	4.0MHz
no loss	2.98	3.42	3.28dBi

Figure 47

Overall, this is a very simple antenna and there are no complications when a typical 50/200 Ohm Balun is used in the inverted configuration. It has achieved excellent SWR bandwidths and excellent RF gains; therefore it should be considered to be one of the "Super-Dipole Antennas".

Diamond 75-80m 12.5-50 Ohm 25'ft H, "Series-section Matching"

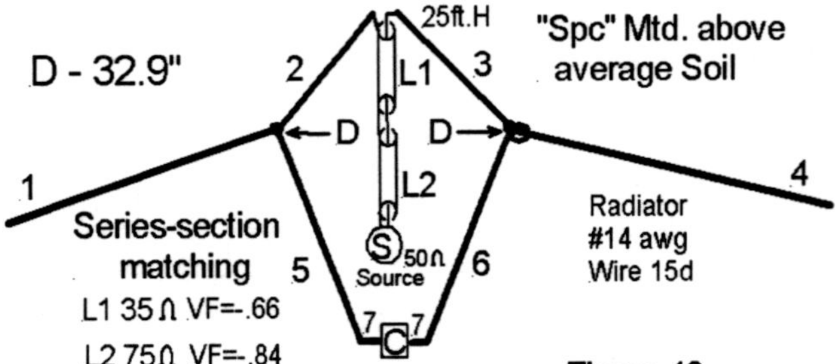

Figure 48

Several Diamond Antennas will be shown mounted quite low at 25' feet above ground. This lower height was chosen to demonstrate the versatility of the various Diamond configurations and secondarily to provide a source for antennas for the average Amateur who may not have access or be able to afford the taller Towers or Masts. All of these will be antennas that have a feed point impedance of 12.5 Ohms and different matching means will be used to achieve a 50 Ohm Source.

The prime reason for modeling antennas with a 12.5 Ohm Source is that they seem to work the best when using a Diamond configuration. More ideal, is to have an antenna that will directly match 50 Ohms and this would eliminate having to use coaxial cable and matching lines, which have standing waves, which will produce RF losses.

When attempting to obtain a higher Source impedance, the common coupling Wires 2 and 3 will change somewhat in length but to obtain a 50 Ohm source the resonant circuit Wires 5 and 6 have to be made exceedingly long, to the extent that the configuration becomes more difficult to model and the shape becomes awkward. During this process, the Capacitor used in this resonant circuit has to be decreased to a very large extent and configurations that differ from the Diamond will have to be used instead. The first 25' foot high Diamond Antenna, as shown in Figure 48 above, uses a Capacitor value of 560pF, which is mounted on Wire 7 and hangs down from the Broad-Banding circuit which consists of Wires 2, 3, 5, 6 and 7, while the 12.5 Ohm Feed-point is located at the Apex of the antenna.

The radiator Wires 1 and 4 have a slope of 15° and extend down to a height of 7.27' feet above ground and out at a distance of 61.83' feet from the center, well out of reach.

This antenna uses "Series-section" Broad-Banding lines in order to match the 12.5 Ohms to a Source impedance of 50 Ohms, which consists of line L1, which is 38' feet of 35 Ohm coax, while line L2 consists of 47' feet of 75 Ohm coax. This makes its total line length 85' feet long, while the antenna is mounted at a height of only 25' feet. This probably will result in total line lengths longer than necessary and with additional RF losses, thereby reducing the Far-field RF gains. This long line is used for both the matching and transmission lines combined.

At a mounting height of 25' feet, this antenna has an SWR averaging less than 2/1 across the major portion of the bandwidth but deteriorates above to approximately 2.5/1 at the band edges as shown in Figure 49. At this height, the antenna slope is 15°. Because of this low angle of radiator slope, it is possible to obtain SWR values below 2/1, whereas, the same antenna mounted at 80' feet and having a steep radiator slope of 35° or 45°, will have difficulties producing an SWR better than this.

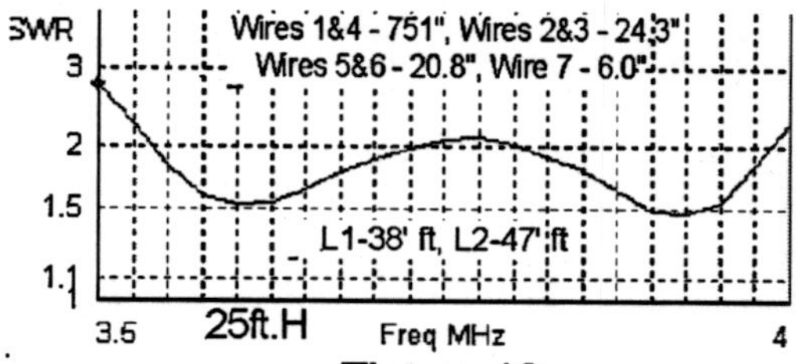

Figure 49

The Far-field RF radiation chart of Figure 50 shows gains of 0.51, 1.43 and 1.12dBi across the 3.5 to 4MHz bandwidth. These radiation gains certainly are much lower than that achieved when the antenna is mounted at a higher height but these figures seem to be what might be expected at this height. The RF losses of 0.64, 0.31 and 0.78dB are higher than desired but because of the "Series-section Matching" coaxial cable lengths of 38' and 47' feet, the losses could be expected to be higher, as shown.

Far-field Gain & Losses at 89.7°			
25ft.H	3.50	3.75	4.0MHz
No loss	1.15	1.74	1.90dBi
w/ loss	0.51	1.43	1.12dBi
RF Loss	0.64	0.31	0.7886dB
Coax-0.5dB/100'/10MHz			

Figure 50

These coaxial cable lengths are exceedingly long, considering that the antenna is mounted at 25' feet above ground and there will generally be excessive coaxial cable used. Other configurations to be presented will have more reasonable lengths but this antenna configuration was shown to make it known that "Series-section Matching" is not always practical, depending on the particular impedances that are being used. It is extremely good for matching a 75 Ohm feed to 50 Ohms, but in matching 25 Ohms to 50 Ohms, the cable lengths become excessive.

Diamond 75-80m 12.5-50 Ohm 25'ft H, with 1/4wL Coax Line

The previous antenna configuration was used but the 12.5 Ohm Feed-point will be matched to 50 Ohms using a 1/4 wL, 25 Ohm matching line, which consists of two 50 Ohm coaxial cables used in a parallel configuration, which results in an equivalent 25 Ohm line. This line is connected between the 12.5 Ohm Feed-point and the 50 Ohm Source. The resulting coaxial line lengths are now 54' feet in length, which is far shorter than the previous antenna, which used "Series-section Matching", as shown in Figure 51.

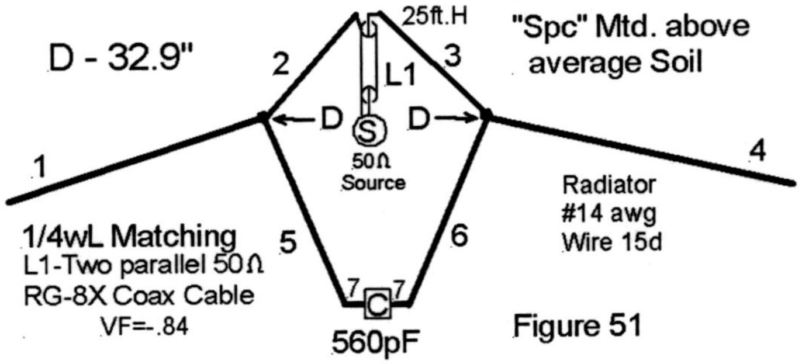

Figure 51

Since this antenna is basically the same as the previous antenna and the only changes are the matching line, with its shorter length, it has substantially the same SWR plot, as shown in Figure 52.

The revised matching line has lower RF losses, as shown in the Far-field RF gain and losses chart of Figure 53. The matching line besides its matching function, being shorter, it has less SWR on the line itself, which reduces the losses further to values of 0.45, 0.14 and 0.44dB across the entire bandwidth. These losses are relatively low, considering the fact that this line service has a 54' foot transmission line, in addition to its matching function.

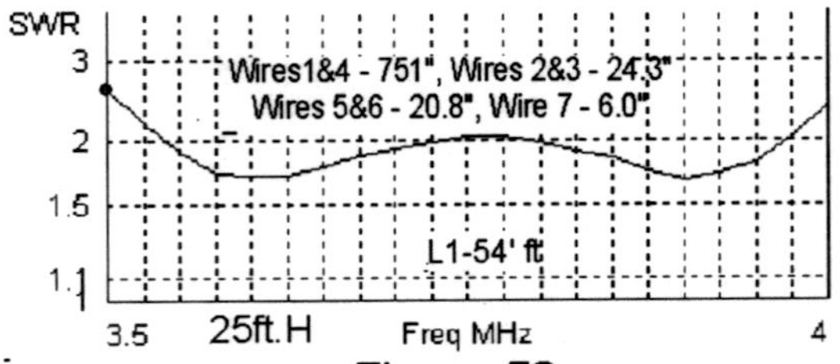

Figure 52

In order to show the effects of adding either the "Series-section Matching Line", or the 25 Ohm 1/4wL coaxial line, the line was removed and the SWR was plotted using a 25 Ohm Source and the resulting SWR plot of Figure 54 is the result. The SWR is relatively flat with an SWR of 2/1 across the bandwidth. When either of the two types of matching lines is used, the resulting SWR averages almost the same but the added matching lines cause less of a flat SWR and develop a greater SWR fluctuation across the bandwidth.

Far-field Gain & Losses at 89.7d.			
25ft.H	3.5	3.75	4.0MHz
No loss	1.16	1.74	1.89dBi
w/ loss	0.71	1.60	1.45dBi
RF Loss	0.45	0.14	0.4486dB
Coax-Two-each 1.0dB/100'/10MHz			

Figure 53

Overall, this antenna when using the 1/4wL matching line has relatively low coaxial losses and also RF gains that are in-line with typical expected gains for this antenna height of 25' feet. These SWR figures are well within the range that internal Antenna-Tuners can properly improve any mismatch.

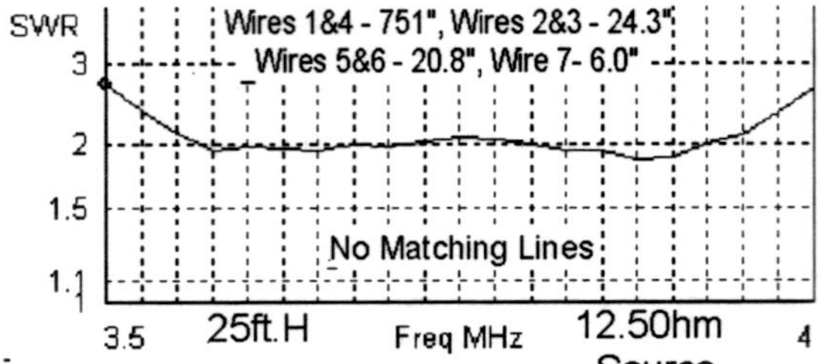

Figure 54

Diamond 75-80m 12.5-50 Ohm 25'ft H, 12.5/50 Ohm Balun

Matching with a 1/4wL line or a "Series-section" coaxial line has inherent coaxial losses, but either of these lines can be eliminated and in place of them a Balun can be used that will match the 12.5 Ohm Feed to a 50 Ohm Source. This will remove all of the coaxial losses but it will introduce some minor loss due to the Balun internal lines.

The Balun, as used with this antenna must be used in a reverse direction from its normal use. A standard 50/200 Ohm Balun can be used and it will have very little degradation. Such an antenna is shown in Figure 55 and as you can see it uses a 560pF Capacitor located on Wire 7, which resonates the Diamond Wires 2, 3, 5, 6 and 7 to the center of the bandwidth and the Balun is placed at the Apex of the antenna, at which point, a good match to 50 Ohms is achieved at the Source.

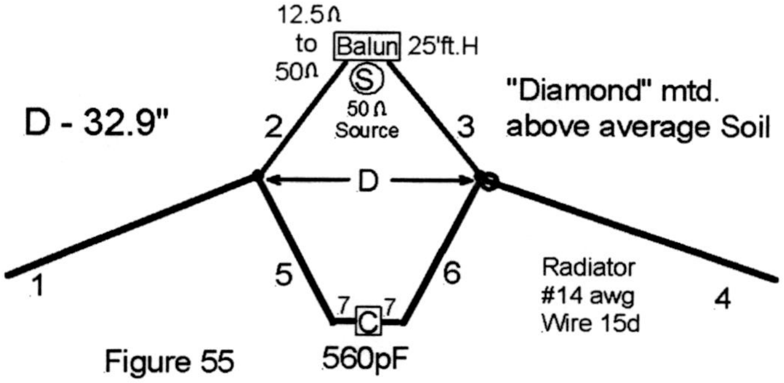

Figure 55

The SWR achieved when this antenna is mounted at 25' feet is shown in Figure 56 and it averages slightly above a value of 2/1. This is slightly higher than the two previous antennas but keep in mind that they had considerably higher coaxial RF losses, which this antenna avoids.

Since there are no coaxial cables used, no coaxial losses can be stated for this antenna and only the Far-field RF gains are shown in Figure 57. The RF gains have progressively increased in small amounts, as the matching lines are changed from "Series-section Lines", to a 1/4wL line and then to the point on this antenna where these lines were eliminated. At a height of 25' feet above ground, the RF gains of Figure 57 are quite good and none of the RF power is dissipated in the coaxial cable but instead it is radiated.

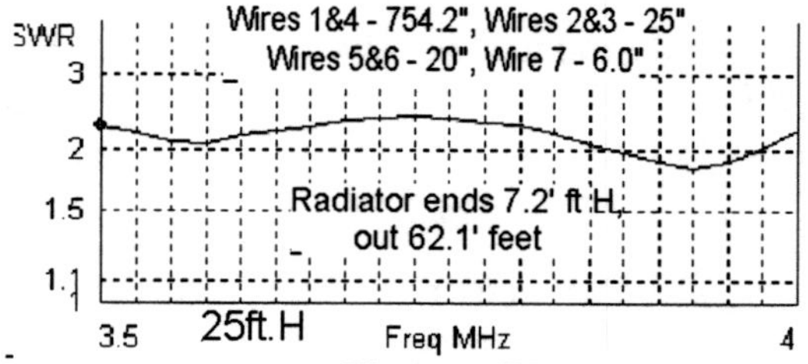

Figure 56

This antenna is simple and inexpensive to build and it should perform extremely well at a height of 25' feet. Most Amateur radio operators will quickly recognize the simplicity and the easy means of having operation on the 75-80m band when building and using this antenna.

Far-field RF Gains at 89.7°			
25ft.H	3.5	3.75	4.0MHz
no loss	0.97	1.61	1.65dBi
Using 50' of 50 Ohm transmission Coax line-loss 0.5dB/100'/10 MHz			

Figure 57

Diamond 160m 12.5-50 Ohm 25'ft H, 12.5/50 Ohm Balun

The previous antenna was scaled from the 75-80m band to 160m and was mounted at a height of 65' feet above ground. At this height, the radiator slope had to be changed to 27° in order to prevent the radiator ends from touching ground. At this new radiator slope of 27°, the radiator ends will be 7.0' feet above ground and extending out from the antenna center to a distance of 112.1' feet.

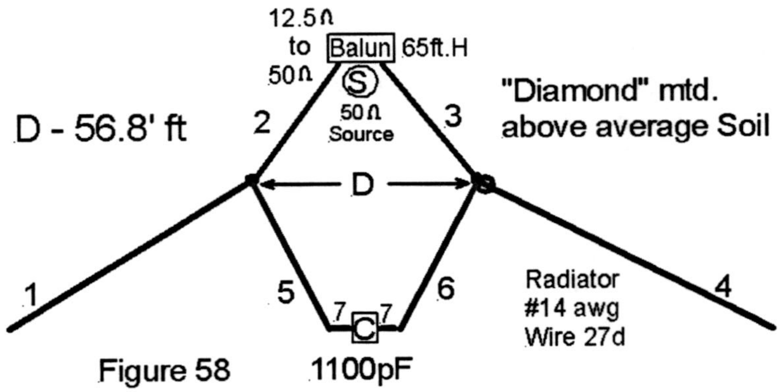

Figure 58

In scaling the antenna to the 160m band, the capacitor value had to be changed to 1100pF in order to have the Diamond configuration resonate to the 160m band center, as can be seen in Figure 58.

The SWR plot of Figure 59 shows that the SWR is only slightly over 2/1 across the entire bandwidth from 1.8 to 2MHz. This is a satisfactory SWR to work with and most internal Antenna-Tuners will improve this. Having this antenna at a height of 65' feet in the 160m band should give an equivalent SWR if it was scaled to the low 75-80m band, since it will then be mounted at 1/2 of the height, which is 32.5' feet.

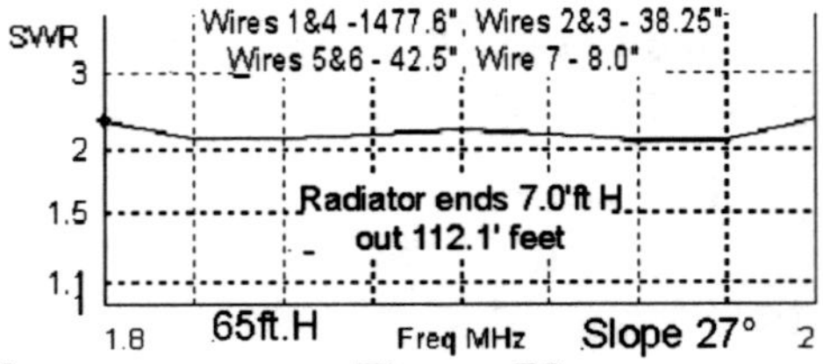

Figure 59

The Far-field radiation gains as shown in Figure 60 are consistent across the entire bandwidth at 2.08, 2.63 and 2.32dBi from 1.8 to 2MHz and these gains are obtained without any coaxial losses. Nevertheless, there will be moderate losses in whatever transmission line length is used.

Far-field RF Gains at 89.7°			
65ft.H	1.8	1.9	2.0MHz
RF gain	2.08	2.63	2.32dBi
Using 70' of 50 Ohm transmission Coax line-loss 0.5dB/100'/10 MHz			

Figure 60

When this antenna is reduced in height from 65' down to 41.5' feet, the radiator slope must be changed to 23° and at that angle the radiator ends will be 7.78' feet above ground and extend outward to 121.4' feet, as shown in the SWR plot of Figure 61.

The SWR plot, as shown in Figure 61, is almost identical to the plot when the antenna was mounted at 65' feet. As the antenna was lowered, the antenna radiator slope was reduced and the reduction is the reason that this antenna can still maintain an SWR of slightly over 2/1.

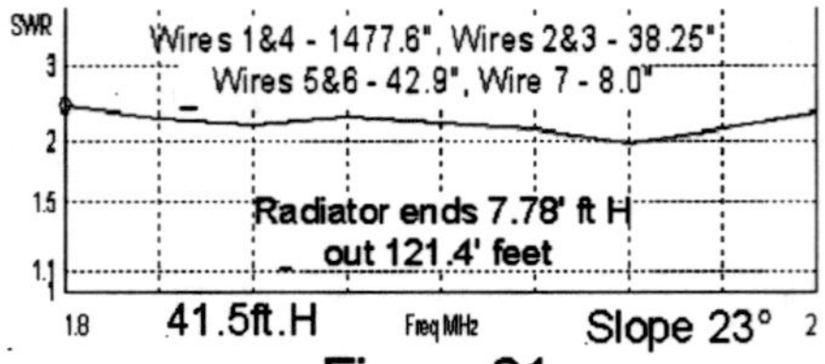

Figure 61

Far-field RF Gains at 89.8°			
41.5ft.H	1.8	1.9	2.0MHz
RF gain	0.28	1.04	0.76dBi
Using 50' of 50 Ohm transmission			
Coax line-loss 0.5dB/100'/10 MHz			

Figure 62

The Far-field RF gains, as shown in the chart of Figure 62 have dropped quite a bit, to values of 0.28, 1.04 and 0.76dBi across the entire 1.8 to 2MHz bandwidth. This drop again appears to be normal, nevertheless, this antenna SWR can be improved with the typical transceiver internal Antenna-Tuner.

This is a relatively simple antenna and the typical 50/200 Ohm Balun can be used in a reverse direction and the performance will not suffer, additionally this Balun will have a minimal of RF loss. This antenna has no other components that can create RF loss except for the radiator wires themselves. Therefore, this is an extremely efficient antenna which should perform well.

Diamond 75-80m 450 Ohm Source without Coax lines

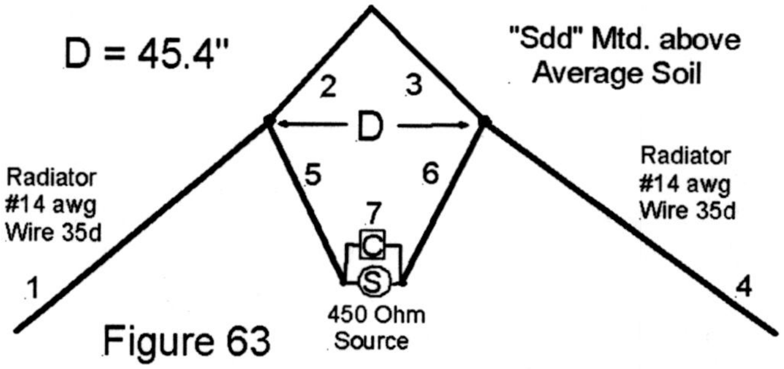

Figure 63

In order to avoid losses in the coaxial cable that might be used for matching, or transmission lines, the Diamond configuration was used and modified, with the Source located on Wire 7. The 405pF resonating Capacitor is placed in parallel with this Source and the impedance is now 450 Ohms. Under these conditions, open wire Ladder/Window-Line with an impedance of 450 Ohms can now be used as the transmission line and there should be no Standing Waves on this line, which might cause additional losses.

This configuration is shown in Figure 63, the antenna is mounted at 60' feet and it has a radiator slope of 35°. The dimension "D" is at a distance of 45.4" inches and an insulated restraining line of that length should be used between these two points in order to maintain the Diamond size.

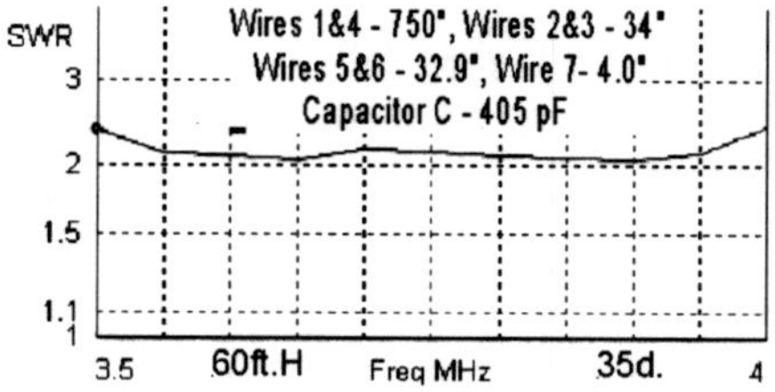

Figure 64

Using this simple configuration, the SWR plot as shown in Figure 64, has achieved an SWR slightly above 2/1, this is very typical of similar antennas in the 75-80m band. This antenna is probably as simple as you might be able to find and still have a very good SWR.

Since this antenna has no coaxial cables in its configuration, no coaxial losses are shown in the Far-field RF radiation chart of Figure 65. Instead, only the Far-field RF gains are shown as 4.90, 5.12 and 5.07dBi across the 3.5 to 4MHz bandwidth. These are satisfactory losses for an antenna mounted at 60' feet with a radiator slope of 35°.

Far-field Gains			
60' ft.H	3.5	3.75	4.0MHz
RF gain	4.90	5.12	5.07dBi

Figure 65

This is a very efficient antenna, since it has no coaxial cable to create RF losses; it is extremely simple in its configuration and is simple to build. A good indicator of the efficiency is to judge whether or not the RF gain is consistent across the entire bandwidth and this particular antenna has maximum gain at 3.75MHz and its drop in RF gain at the band edges are very minimal, making this a very efficient antenna.

The transmission line would normally consist of the 450 Ohm Ladder/Window-Line, which has an RF loss of 0.06dB/100'/10MHz, making this an extremely low loss line. The biggest difficulty with using such a line is to find a means of matching it to 50 Ohms at the Transmitter or Transceiver end.

If this matching to 50 Ohms cannot easily be achieved, the following antennas can provide a means of matching the 450 Ohm Source to 50 Ohms and not use the Ladder/Window-Line.

Diamond 75-80m 450 Ohm, 300 Ohm line and a Balun

One method of matching a 450 Ohm Source to 50 Ohms is to use a 300 Ohm 1/4wL matching line to match the 450 Ohm Source to 200 Ohms and then using a typical 50-200 Ohm Balun to transform the impedance down to a 50 Ohm Source. The 300 Ohm line can be a 300 Ohm Twin-lead open-wire line, as will be described in Figure 66.

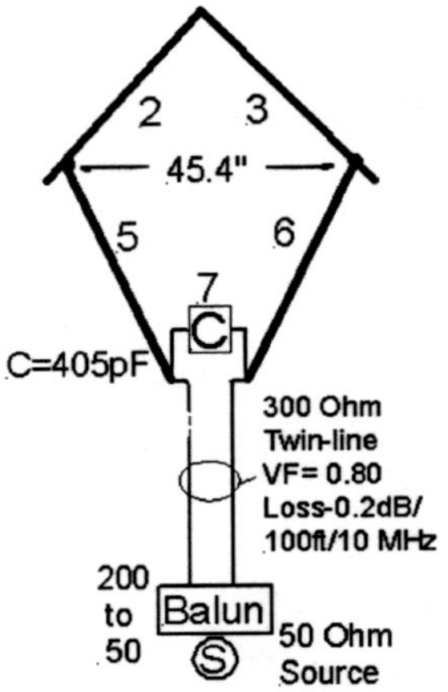

Figure 66

The antenna as shown in Figure 66 is in an abbreviated form, with the major portion of the radiator wires not shown. As you can see the 300 Ohm Twin-line is connected to the antenna Source and the 405pF resonating Capacitor is still placed in parallel with this line. From the Twin-line description, this line has a VF= 0.8 and the specified losses are 0.2dB/100ft'/10MHz. This is a considerably greater loss than the 450 Ohm Ladder/Window-Line but the matching is only required between fairly close impedance values of 450 and 200 Ohms and therefore the RF losses will be very moderate.

Figure 67

The SWR plot for this antenna, as shown in Figure 67, has dropped very slightly, probably due to the Twin-line RF losses and is very good at approximately an SWR of 2/1.

By having the 300 Ohm Twin-line, this antenna now has some RF losses attributable to this line but the Far-field RF (Twin-line) losses due to this line are only 0.14, 0.06 and 0.13dB, which are exceedingly low. Since this line is 56.2' feet long, this forms part of the transmission line length running towards the transmitter or transceiver. An equal length of this line if it was to be used for a transmission line, without any standing waves, would likely have a loss of approximately 0.1dB and if you subtracted this figure from the stated losses in Figure 68, the losses would dramatically decrease to values that are rather insignificant.

Far-field Gains			
60ft.H	3.5	3.75	4.0MHz
No Loss	4.93	5.13	5.06dBi
With Loss	4.79	5.07	4.93dBi
Coax loss	0.14	0.06	0.13dB

Figure 68

This Diamond Antenna is very simple to understand and to build and it has extremely low losses, making it one of the more efficient antennas. The Diamond configuration has given the capabilities to be able to adjust the degree of RF coupling between the Diamond resonant circuit and the radiator. This permits optimizing the shape of the SWR plot, while the lengths of Wires 5 and 6 should be adjusted to maintain resonance at the band center with various tolerances of the 405pF Capacitor. When having all of these capabilities, this antenna should be classified as a "Super-Dipole Antenna".

Diamond 75-80m 450 Ohm, 150 Ohm line and a Balun

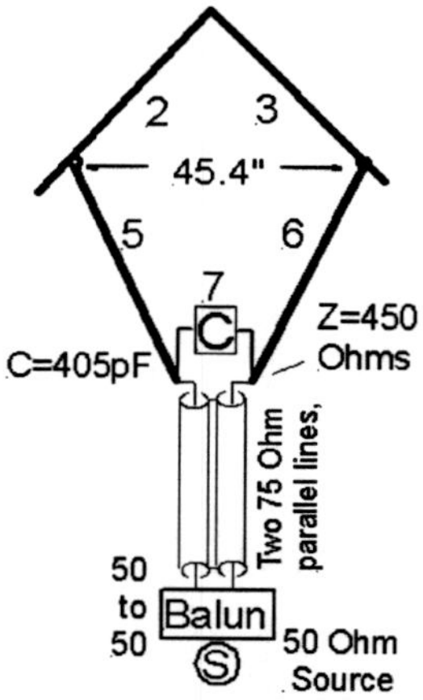

Figure 69

Although the use of coaxial cable in the various antennas has been generally avoided because of the typical Far-field RF radiation gain and losses, there are times where its use can be justified when used in certain configurations and this antenna is one of those cases.

The previous antenna, which used a 300 Ohm Twin-line, was modified to use coaxial cable for each of the wire lines, as shown in Figure 69. Each of the coaxial cables is changed to an impedance of 75 Ohms and their braids are connected together forming a balanced 150 Ohm line. This combined line will transform the 450 Ohm antenna to an impedance of 50 Ohms. This new Source is balanced and therefore a 50/50 Balun must be used at the Source. No other changes have been made to the antenna configuration and these changes are simple. (Remember also, that as in Fig. 66 the major portions of the radiators in Fig. 69 are also not shown).

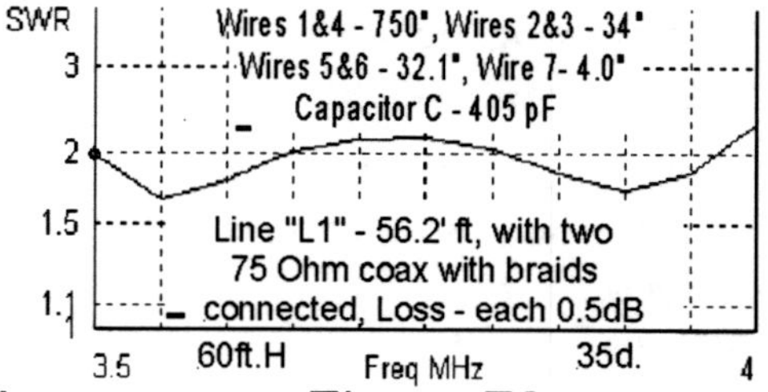

Figure 70

The 75 Ohm coaxial cables used have a Velocity Factor of 0.84 and the specified losses are 0.5dB/100'/10MHz. The resulting balanced line will have a loss of 1/2 or 0.5dB and the total line length is 56.2' feet. With the antenna mounted at 60' feet in height, this seems like a very suitable length.

With this new arrangement, it produced an SWR plot averaging below 2/1 across the entire 3.5 to 4MHz bandwidth, as shown in Figure 70.

Far-field Gains			
60ft.H	3.5	3.75	4.0MHz
No Loss	4.93	5.13	5.06dBi
With Loss	4.65	5.05	4.75dBi
Coax loss	0.28	0.08	0.31dB

Figure 71

Even though coaxial cable with a 1/4wL line length of 56.2' feet was used, surprisingly low losses are shown in the Far-field RF gain chart of Figure 71 with losses at 0.28, 0.08 and 0.31dB. When you consider that this 150 Ohm line is also part of the transmission line, such a line would have RF losses of possibly 0.1dB, which you might add to the Far-field RF gains, this would result in an extremely low loss antenna with extremely high efficiency.

The Far-field RF gains as shown in Figure 71 have changed only very slightly from that of the previous antenna and a slight reduction is shown.

Even though coaxial cable has been used, the RF losses due to this cable are minimal and this antenna has all of the qualities of the previous antenna and it should be classified as a "Super-Dipole Antenna".

This concludes Chapter 2, which provided antennas that utilize a discrete fixed Capacitor in order to resonate the Diamond configuration. In order to avoid this Capacitor and its tolerance considerations, the following chapters will include antennas to where the Capacitor is deleted and a capacitive coaxial stub is used instead.

Super Max-Dipole Antennas

By: Edward J. Shortridge W4JOQ

Chapter 3

Super-Delta Horizontal Antennas

Introduction

In early chapters of this book and the previous book titled "Super-Dipole Antennas", it had been clearly demonstrated that basic dipole antennas that did not include Broad-Banding were not practical for use in the 75-80m or 160m radio Amateur bands. Unfortunately, these basic antennas had an SWR bandwidth of 1.5/1 that covered only a fraction of the total bandwidth, making it necessary to use an external Antenna-Tuner which had to be readjusted constantly as the operating frequency is changed, especially towards the band edges which might have SWR values as high as 5 to 10/1.

With this in mind, it was necessary to develop antennas that were Broad-Banded sufficiently to the point of having an SWR well below 3/1 which most internal Antenna-Tuners will affectively be able to improve the SWR. The vast majority of the modeled Broad-Banded antennas to be presented are able to achieve SWR values in the range of 2/1 and the few exceptions were able to achieve considerably better figures.

Having a low SWR is only one of many necessary characteristics that identify Super-Dipole Antennas and here is a list of requirements that seemed quite necessary to earn this title.

- Can readily be adjusted to a 50 Ohm source
- Adjustable Broad-Banding coupling provided
- Fixed capacitor or capacitive stubs can be used
- 2/1 SWR values obtained across the 75-80m or 160m bands
- Low RF voltages on capacitor or capacitive stubs
- Any capacitive stubs used are very short
- Simple to build and mount
- Highest efficiency for Broad-Banded antennas

Two antennas that were developed in Western North Carolina within view of High Rock Mountain, meet all of these requirements and to give them distinctive names, I have taken the liberty to call them the "High Rock Super-Delta Horizontal" and the "High Rock Super-Diamond Inverted V" antennas. This chapter will be devoted entirely to the Super-Delta Horizontal and then Chapter 4 will cover the Super-Diamond.

75-80m Super-Delta Horizontal with 2 Fixed capacitors

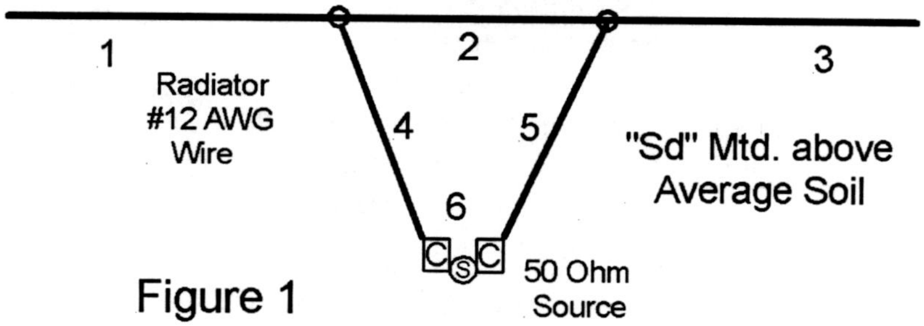

Figure 1

The first antenna to be presented is the Super-Delta Horizontal, which uses two Fixed Capacitors in order to resonate the Broad-Banding resonant circuit to the center of the 75-80m bandwidth as shown in Figure 1. The 50 Ohm Source is located at the juncture of these two Capacitors which are in series with each other.

This particular antenna was presented in articles in the AntenneX On-line Antenna Magazine N0. 160 and 161 dated August and September 2010. In this chapter we will provide many more versions with updated information not previously available.

This antenna uses a horizontal radiator consisting of Wires 1, 2 and 3. A Delta shaped resonant circuit, consisting of Wires 4, 5 and 6 are used in a resonant circuit that requires the two series Capacitors C to resonate near the center frequency of the 75-80m band.

The common Wire 2 can be adjusted for a variety of coupling values in order to reach a point of Critical-Coupling, at which point Broad-Banding will begin to occur. This coupling can then be adjusted to give the best SWR across the entire bandwidth.

The Delta wire lengths are adjusted along with the Capacitor values in order to maintain a triangular shape, with Wires 2, 4 and 5 at approximately the same length in order to avoid undesirably narrow angles.

Two 330pF Capacitors are used primarily to achieve a balanced Source, which is located between these two Capacitors. The use of a balanced 50/50 Balun located at the Source is recommended.

Wire 6, which contains the two 330pF Capacitors and the Source, can be adjusted upward or downward in order to re-resonate the Delta, so as to compensate for the tolerances of these 2 Capacitors. Please note that this antenna uses a larger wire size of #12 instead of #14 AWG and this larger wire size has helped to provide an extremely good SWR as shown in Figure 2, when this antenna is mounted at a height of 80' feet. Most other Super-Delta horizontal antennas cannot meet SWR values below the 1.5/1 as provided by this antenna. Other than the wire, the only other components are the two series connected Capacitors. It is commonly known that well designed Capacitors may have Q values as high as 5000 and if a good transmitting type Capacitor that has high current and voltage ratings are used, they will not contribute to any loss of the Far-field radiation gain.

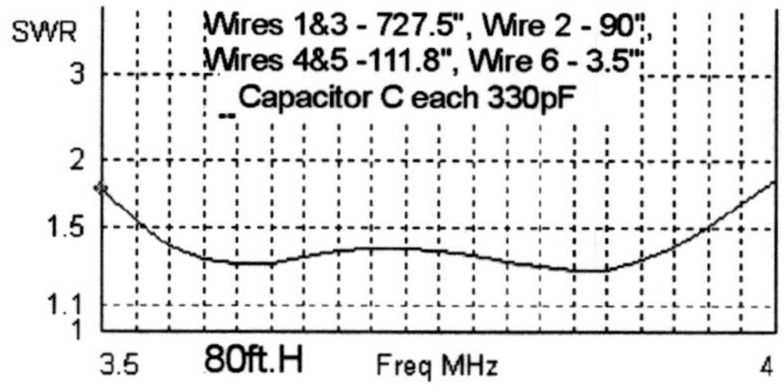

Figure 2

When this antenna is lowered to a height of 60' feet above Average-soil and the radiator lengths are shortened to once again establish resonance in the center of the band, a very good SWR value of 1.5/1 is obtained across the center portion of the bandwidth, however it deteriorates to approximately 2/1 at the band edges, as shown in Figure 3. In contrast, most other Super-Delta horizontal antennas generally can only provide an SWR in the vicinity of 2/1.

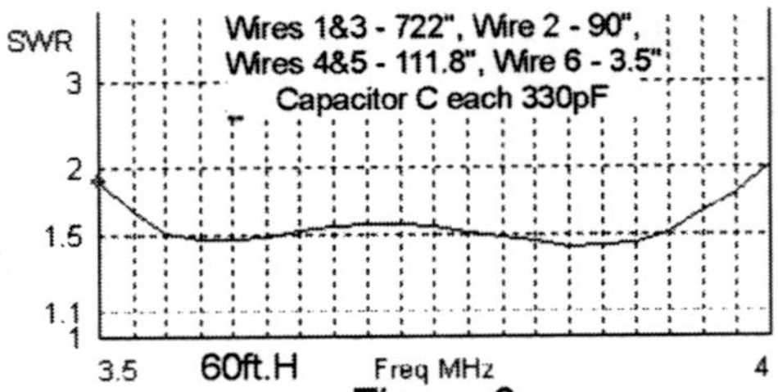

Figure 3

Far-field Gain at 73.1° 60' ft. H			
No Coax	3.5	3.75	4.0MHz
losses	6.13	6.22	6.12dBi
Two 330pF series Capacitors			

Figure 4

This antenna has very good Far-field RF gains of 6.13, 6.22 and 6.12dBi across the entire bandwidth from 3.5 to 4MHz, as shown in Figure 4. These are our very typical gains for this antenna height of 60' feet; however the SWR figures are extremely good.

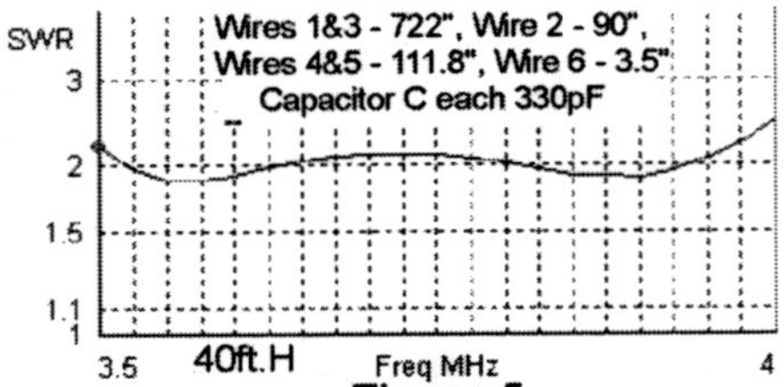

Figure 5

When this antenna is lowered from 60' down to 40' feet, the SWR has deteriorated to a value of approximately 2/1 across the band center and only slightly higher at the band edges, as shown in Figure 5.

This antenna has extremely good characteristics; however it requires #12 wire and the fact that 2 Capacitors of equal values and tolerances are used to maintain a balanced feed. For these reasons newer versions of the Super-Delta antenna will be presented.

75-80m Super-Delta Horizontal with 2 short open-ended Stubs

This antenna has substantially the same configuration as the previous antenna and still uses #12 AWG wire. The two 330pF Capacitors that were located on each side of the Source have been eliminated and as a substitute, two Capacitive Short-line open-ended Stubs are used instead. These two 119" inch long Stubs are then mounted at the upper end of the Delta attached to the main radiator, as shown in Figure 6.

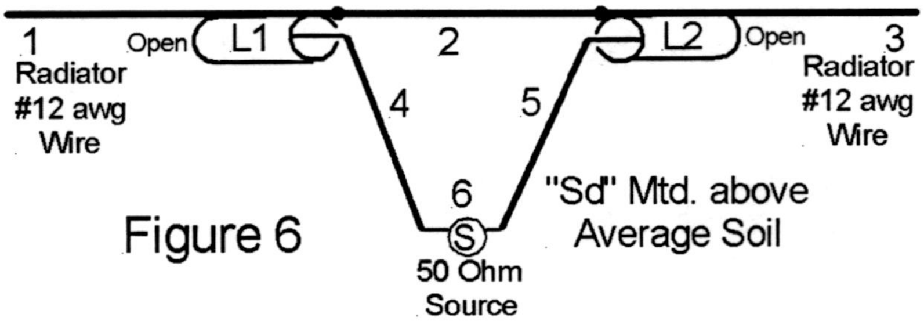

Figure 6

After these stubs are mounted as shown in Figure 6, the source, which is located on Wire 6 can be moved upward or downward to obtain the lowest symmetrical SWR bandwidth across the entire 3.5 to 4MHz band.

To help facilitate this change in the wire lengths of Wires 4 and 5, it is wise to provide wire clamps at the Source, mounted in a manner so that Wires 4 and 5 readily slide through this clamp and then are tightened securely. In order to do this there should be extra several inches of extensions to Wires 4 and 5 to allow these wires to be increased in length.

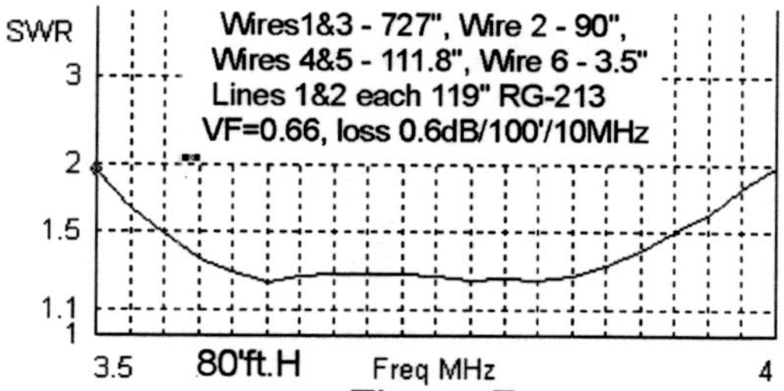

Figure 7

After adjusting for the best SWR bandwidth, when mounted at 80' feet, the SWR is substantially lower than 1.5/1 across the major portion of the bandwidth and deteriorates to 2/1 at the band edges, as shown in Figure 7.

This is an extremely good SWR and when the antenna is lowered to 60' feet; it still remains slightly below 1.5/1 over the major portion of the bandwidth, as shown in Figure 8.

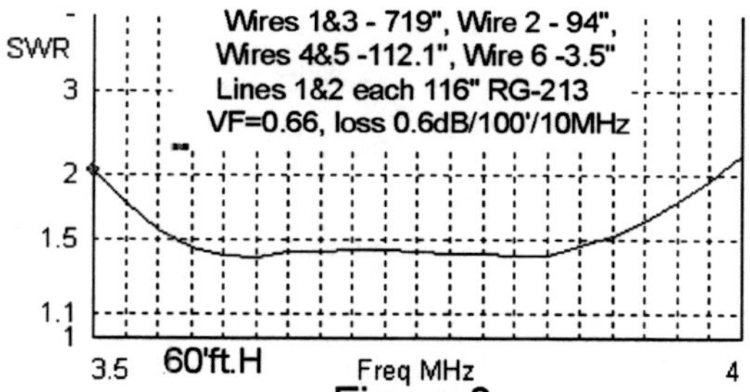

Figure 8

Often times, when the SWR is extremely low, it is necessary to check the Far-field RF gain and loss chart, such as that shown in Figure 9. In this chart all of the RF gain figures are above 6 dBi, which is very good. And the coaxial losses are extremely low at, 0.04, 0.03 and 0.05dB indicating that this antenna is extremely efficient, has good radiation gains and can handle extremely high RF powers due to the two Stubs being in series, which doubles the specified break-down voltage rating.

Far-field Gain & Losses at 73.1°			
60ft.H	3.50	3.75	4.0MHz
No loss	6.12	6.19	6.08dBi
w/ loss	6.08	6.17	6.03dBi
RF Loss	0.04	0.03	0.05dB
Coax loss - RG-213 Vf=0.66			

Figure 9

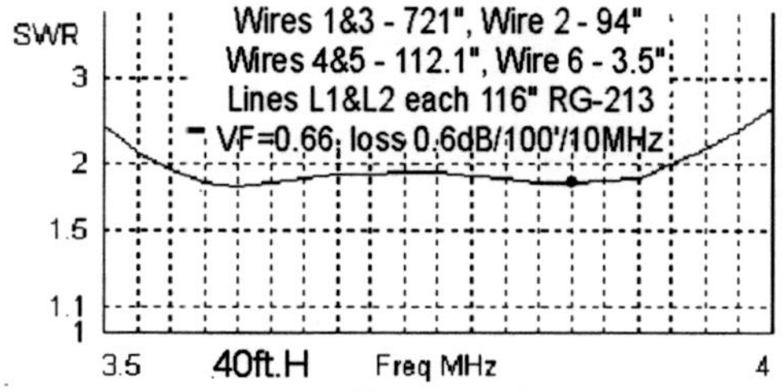

Figure 10

Lowering the antenna from 60' feet down to 40' feet caused a distinct deterioration of the SWR from 1.5/1 up to 2/1 across the entire 3.5 to 4MHz bandwidth, as shown in Figure 10. This is still an unusually low SWR for this height of 40' feet and this antenna has all the qualities to classify it as a "Super-Dipole Antenna". This antenna has unusually good characteristics and should perform extremely well.

This antenna and the previous one shown in Figure 1 are antennas that were developed and presented quite a few years earlier, whereas the following antennas which present new configurations are the newest to be presented to date.

75-80m Super-Delta Horizontal with a single Fixed Capacitor

This antenna has a different newly developed configuration, which uses a single Fixed Capacitor at the bottom of the Delta on Wire 6 and has the Source centered on Wire 2. By having the Source at this point, permits the use of a single Capacitor on Wire 6, as shown in Figure 11.

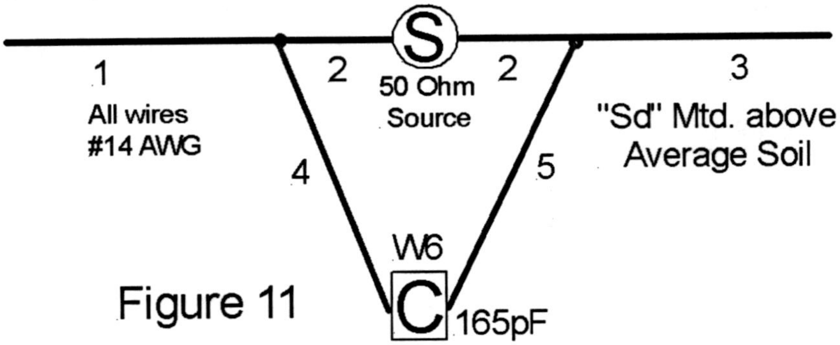

Figure 11

The antenna was mounted at a height of 85' feet and adjusted for the best SWR and the results, as shown in Figure 12 are quite good and slightly over an SWR of 1.5/1 and only slightly higher at the band edges. Even though the Source is now in the middle of the common coupling Wire 2, instead of Wire 6, the same approximate SWR has been achieved as compared with the previous antenna.

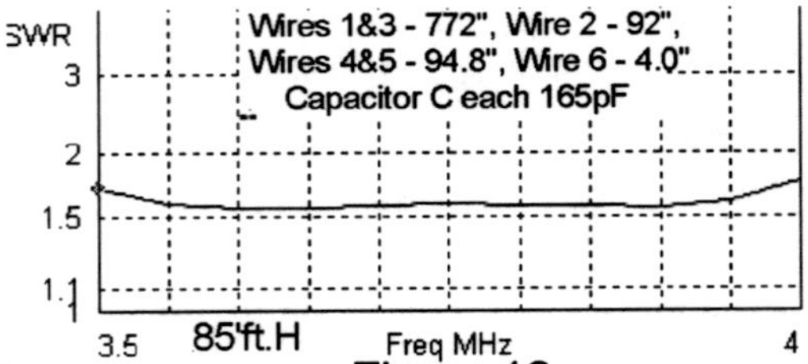

Figure 12

Far-field Gain at 48.4° 85'ft. H			
No Coax	3.5	3.75	4.0MHz
losses	6.06	6.06	6.23dBi
A single 165pF Capacitor			

Figure 13

Since this antenna uses a Fixed Capacitor instead of a Capacitor coaxial Stub, there can be no coaxial losses and the efficiency should be extremely high. Along with the high efficiency, the RF gains as shown in Figure 13 are very good at 6.06, 6.06 and 6.23dBi across the 3.5 to 4MHz bandwidth.

The objective of trying to reduce the use of two Capacitors down to a single one has been achieved and the RF gains are normal, compared to those of other antennas and the efficiency is equally as high. Therefore, we have reduced this antenna to a very simple configuration and the only item of any consequence is the Capacitor, its mounting and its tolerance considerations. Other than the Capacitor, there are no other components necessary to achieve this excellent performance. Lowering the antenna from 85' to 65' feet causes some degradation of the SWR but it is well below 2/1 across the entire bandwidth as can be seen in Figure 14. The RF gains, as shown in Figure 15, are slightly higher than those when the antenna was mounted at 85' feet; however these gains are at a radiation angle of 63.4°, whereas at a height of 85' feet, the maximum radiation was 48.4°.

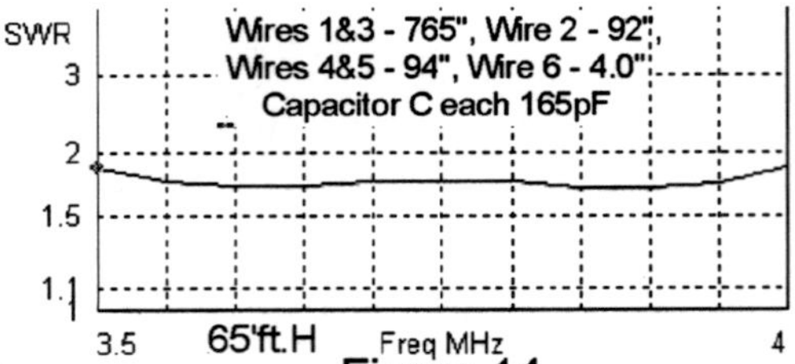

Figure 14

Far-field Gain at 63.4° 65'ft. H			
No Coax	3.5	3.75	4.0MHz
losses	6.23	6.21	6.10dBi
A single 165pF Capacitor			

Figure 15

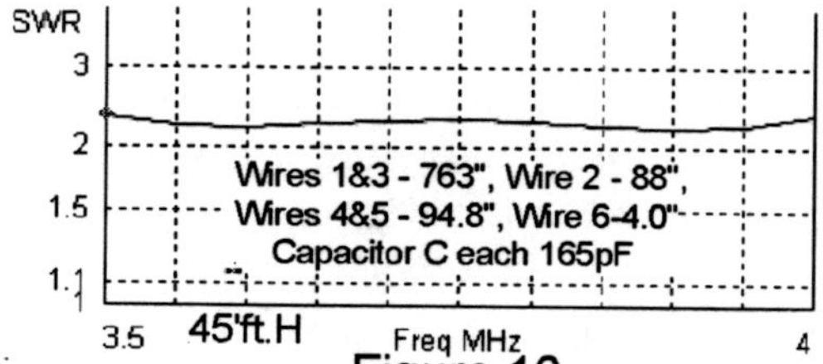

Figure 16

As the antenna is lowered further to a height of 45' feet, the SWR has deteriorated as expected to somewhat above 2/1, however it is flat across the entire 3.5 to 4MHz bandwidth, as shown in Figure 16.

Far-field Gain at 89.9° 45'ft. H			
No Coax	3.5	3.75	4.0MHz
losses	6.37	6.48	6.43dBi
A single 165pF Capacitor			

Figure 17

The Far-field RF gains, as shown in Figure 17, has increased further to 6.37, 6.48 and 6.43dBi across the bandwidth. This might seem odd that the RF gain increases as the antenna is lowered, however you must look at the angle of radiation, which has increased from 48.4° at 85' feet, to an angle of 63.4° at 65' feet and once again increasing to an angle of 89.9° at 45' feet. At this point, if the antenna is lowered further, there cannot be any further increase in the angle of radiation since it is already at 90°.

When the antenna is mounted at elevated heights of 65' and 85' feet, the lower angle of radiation distributes the RF power in all horizontal directions, however as the antenna is lowered, the angle of radiation increases to 90° concentrating the bulk of the RF energy in a vertical beam at 90° and this concentration as the antenna is lowered, causes the increase in RF gain.

Not everyone has the facilities to mount an antenna at 45' feet or higher, therefore we will show this antenna lowered to a height of 25' feet above Average-soil. The SWR plot of Figure 18 shows that the SWR has deteriorated to the point slightly below an SWR of 3/1. Most internal Antenna-Tuners are capable of functioning with this SWR value of 3/1.

As this antenna was lowered from 85' to 65' and then down to 45' feet in height, the radiator lengths had to be decreased to compensate for the added capacity to ground.

However, as the antenna is lowered from 45' feet, down to 25' feet, the reverse has occurred and antenna Wires 1 and 3 had to be increased from 763" inches, when mounted at 45' feet to a length of 769" inches when mounted at 25' feet. This is an unexpected occurrence that cannot be accounted for; nevertheless it is a necessary requirement. When this antenna is mounted at 45' feet, its maximum radiation is at 89.9° and when the antenna is lowered further to 25' feet, as shown in the Far-field RF gain chart of Figure 19, it is still at the high angle approaching 90°. As mentioned previously, lowering the antenna to this height cannot increase the angle of radiation beyond 90°, so there can be no further concentration of RF energy in one direction, so the result is that the RF gains have decreased to 4.55, 4.94 and 5.02dBi, as shown in Figure 19.

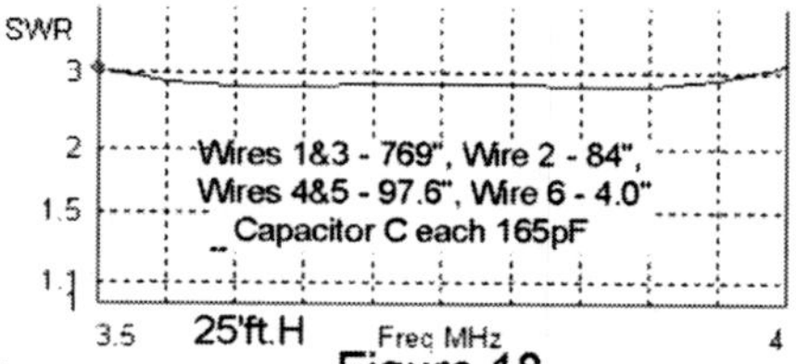

Figure 18

Far-field Gain at 89.8° 25'ft. H			
No Coax	3.5	3.75	4.0MHz
losses	4.55	4.94	5.02dBi
A single 165pF Capacitor			

Figure 19

RF Vrms using 200W RF power			
Frequency	3.5	3.75	4.0MHz
85'ft H	257	257	617 V
25'ft H	205	503	702 V

Figure 20

The RF voltage, as measured across the 165pF Capacitor "C" was measured and calculated for an RF power of 200 Watts, as shown in the chart of Figure 20. This voltage will double when using 800 Watts of RF power. This chart shows the RF voltages across this capacitor at both 85' and 25' feet in height and at frequencies of 3.5, 3.75 and 4MHz.

This High Rock Super-Delta Horizontal Antenna is a simple antenna to understand and build, but other versions will be shown substituting Short-line Capacitive Open-ended Stubs that can replace this Capacitor. The Capacitor value has no means of tweaking or adjusting the Capacitor value slightly, whereas the coaxial stubs can be shortened in discrete steps. If stub lengths are chosen that are slightly longer than specified, then they can be shortened in small increments by snipping off the open end in small increments while noting the change in the SWR bandwidth plot. This allows for adjustment, whereas the Fixed Capacitor does not.

There are other means of compensating for the Fixed Capacitor tolerance that works well and that was explained previously in which the Capacitor housing had clamps, through which the resonant circuit Wires 4 and 5 passed through, permitting adjustments of the Capacitor housing upward or downward slightly to compensate for these Capacitor tolerances and then being able to tighten the clamps onto the downward wires.

75-80m Super-Delta Horizontal with Low Open-end Stubs

Here is another High Rock Super-Delta horizontal antenna that is almost identical to the previous antenna, except that the 165pF Capacitor "C" has been substituted and replaced by Short-line open-ended capacitive Stubs L1 and L2, which are connected in series as shown in Figure 21.

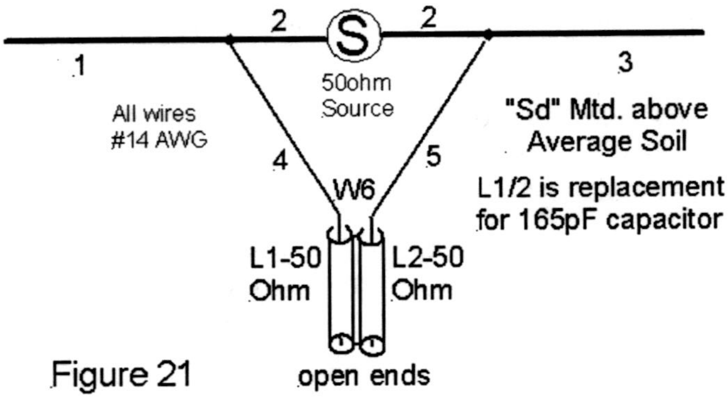

Figure 21

A single coaxial stub could not be used since it is an unbalanced line and because poor SWR and radiation results would occur. Therefore, two 50 Ohm coaxial lines are used and placed in a balanced configuration as shown in Figure 21. The parades of the cables are connected together to achieve this balance.

Whenever the Source is placed on the center of Wire 2, there is nothing normally provided to secure Wire 6 or anything attached to it, such as a resonating Capacitor or coaxial stubs.

Therefore, to prevent this end of the Delta from being whipped around in different shapes by the wind, some means should be provided to hold it in place.

The Source at the center of Wire 2 should utilize a 50/50 Balun and when the 50 Ohm transmission line is connected and run downward these two balanced stubs can be taped or strapped to the outside of the coax and a balanced feed to this antenna will result and be maintained in all weather conditions.

Since the only changes to this antenna, compared with the previous antenna, that uses a single Fixed Capacitor, is that the Capacitor was removed and the two parallel Stubs that are connected in series have been substituted. As a result, the SWR plot, as shown in Figure 22, when mounted at a height of 65' feet, is noticeably the same as the previous antenna mounted at the same height.

Figure 22

The Far-field RF radiation chart of Figure 23 shows that this antenna has substantially the same RF gains across the bandwidth as the previous antenna, resulting in RF coaxial losses that are extremely low at 0.01, 0.02 and 0.06dB across the 3.5 to 4MHz bandwidth. Since the coaxial stubs are relatively short and connected in series, the breakdown voltage for these coaxial cables is extremely high and they produce a very minimal RF loss.

Far-field Gain & Losses at 63.8°			
65ft.H	3.5	3.75	4.0MHz
No loss	6.23	6.21	6.09dBi
w/ loss	6.22	6.19	6.03dBi
RF Loss	0.01	0.02	0.06dB
Coax loss - RG-213 Vf=0.66			

Figure 23

At a mounting height of 45' feet, the SWR plot is almost exactly the same as the previous antenna, as seen in Figure 24, with an SWR of 2/1 across the majority of the bandwidth, additionally the Far-field RF gains at a height of 45' feet are also noticeably the same as the previous antenna and the coaxial losses are quite low at 0.01, 0.05 and 0.09dB as shown in Figure 25.

Figure 24

Far-field Gain & Losses at 89.7°			
45ft.H	3.5	3.75	4.0MHz
No loss	6.38	6.50	6.42dBi
w/ loss	6.37	6.45	6.33dBi
RF Loss	0.01	0.05	0.09dB
Coax loss - RG-213 Vf=0.66			

Figure 25

Lowering the antenna to a height of 25' feet, the SWR plot of Figure 26 and the Far-field RF gains and loss chart of Figure 27 are noticeably the same as the previous antenna mounted at the same height.

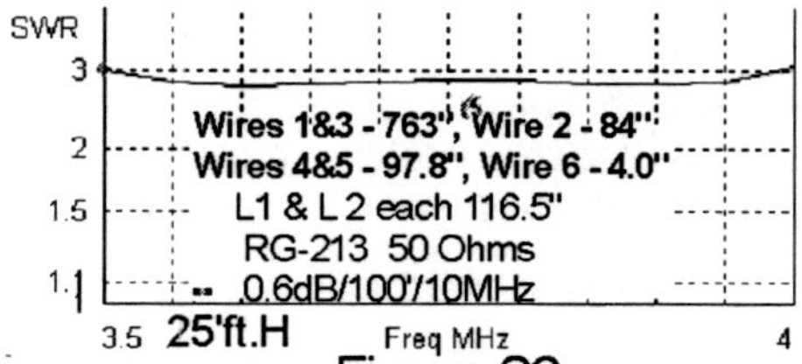

Figure 26

Far-field Gain & Losses at 89.8°			
25ft.H	3.5	3.75	4.0MHz
No loss	4.94	4.94	5.01dBi
w/ loss	4.92	4.92	4.91dBi
RF Loss	0.02	0.02	0.10dB
Coax loss - RG-213 Vf=0.66			

Figure 27

Since the Capacitive Coaxial Stubs are relatively short, the RF voltages on them at heights of 85' and 25' feet as shown in Figure 20 are still valid, even though we have changed from a Fixed Capacitor to capacitive coaxial stubs.

75-80m Super-Delta Horizontal with 2 High Stubs

Instead of having the two capacitive coaxial short-line Stubs mounted at the bottom of the Delta, they can be moved upward and attached to each side of the radiator as shown in Figure 28. These Open-ended stubs will be somewhat the same length as previously used and they will be inserted in the upper ends of Wires 4 and 5; however they should be mounted securely to the radiator Wires 1 and 3. Their lengths and the lengths of Wires 4 and 5 will change somewhat as well.

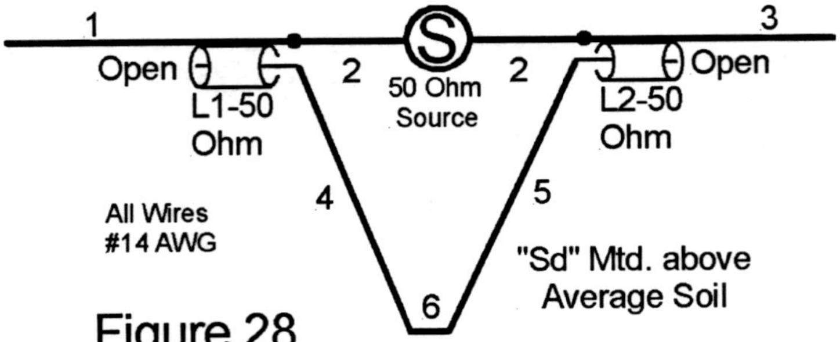

Figure 28

The lengths of these two Stubs L1 and L2 can be kept the same or slightly longer if necessary, while the main adjustment for re-resonating the Delta is achieved by moving Wire 6 downward while attached to Wires 4 and 5, this in effect increases the lengths of these two wires. They should be readjusted to achieve substantially the same SWR plot as was achieved in the previous antenna.

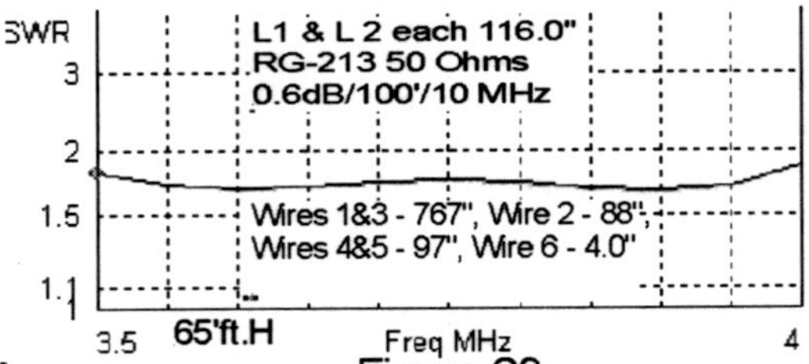

Figure 29

After re-resonating the Delta by this means, the SWR was plotted as shown in Figure 29 and this plot which is for a height of 65' feet is substantially the same as the previous antenna at the same height.

As you can see in Figure 28, Wire 6 does not have a Source, a Capacitor or any coaxial stubs attached to it and therefore it has no weight and unless it is tied in place, it will distort and flop around under windy conditions and the SWR and RF radiation will deteriorate. Therefore, it is necessary to put an insulated stiffener onto Wire 6 and secure it to a downward transmission line that runs from a 50/50 Ohm Balun, which is attached to the Source located on Wire 2.

Far-field Gain & Losses at 63.8°			
65ft.H	3.5	3.75	4.0MHz
No loss	6.38	6.38	6.30dBi
w/ loss	6.37	6.36	6.23dBi
RF Loss	0.01	0.02	0.07dB
Coax loss - RG-213 Vf=0.66			

Figure 30

The Far-field gain and loss chart of Figure 30 shows RF gains of approximately the same as the previous antenna and its RF coaxial losses due to the Stubs is very low at 0.01, 0.02 and 0.07dB, making this a very efficient antenna.

As the antenna is lowered to 45' feet, the radiator Wires 1 and 3 had to be reduced in length and the resulting SWR bandwidth has deteriorated to slightly above an amount of 2/1, as can be seen in Figure 31.

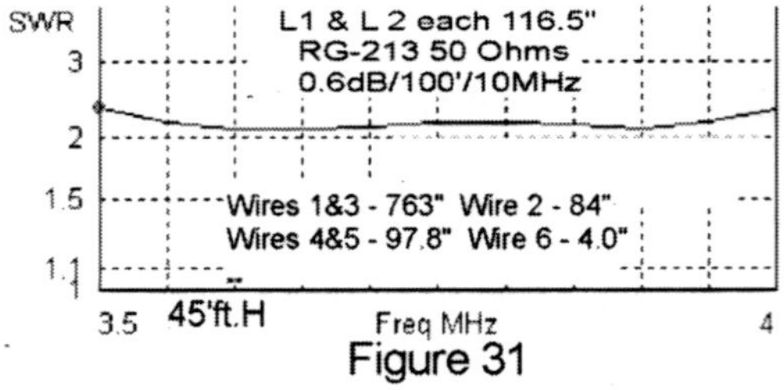

Figure 31

The Far-field RF radiation gains of the chart in Figure 32 shows that the major angle of radiation is now at 89.7°, whereas at 65' feet as shown in Figure 30 the angle was 63.8°. The result is that the gain at 45' feet is slightly higher than the previous height and that this trend was previously indicated. The RF gains are very good and the coaxial losses have actually reduced very slightly to 0.01, 0.01 and 0.07dB across the entire 3.5 to 4MHz band.

Far-field Gain & Losses at 89.7°			
45ft.H	3.5	3.75	4.0MHz
No loss	6.51	6.66	6.62dBi
w/ loss	6.50	6.65	6.55dBi
RF Loss	0.01	0.01	0.07dB
Coax loss - RG-213 Vf=0.66			

Figure 32

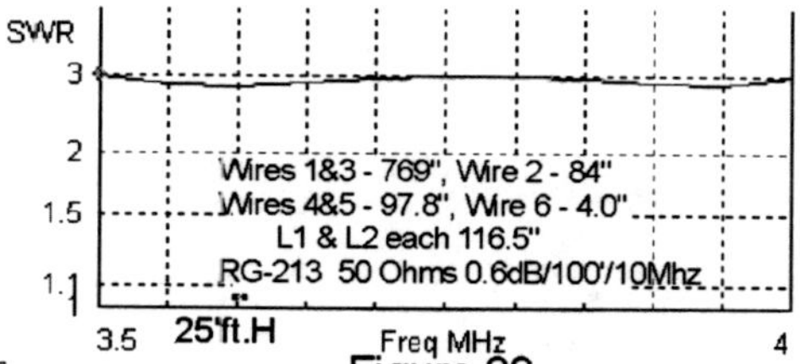

Figure 33

At a height of 25' feet, the SWR has deteriorated further to slightly below 3/1, which is still workable with Internal Antenna-Tuners. Figure 33 also shows that both radiator ends had to be increased in length by 6" inches in order to re-resonate the antenna.

At this lower height of 25' feet, the RF gains have begun to decrease to 4.66, 5.08 and 5.12dBi and the radiation angle is substantially 90° as shown in Figure 34. The coaxial losses have increased very slightly at 4MHz to a value of 0.10dB, which is an insignificant change.

Far-field Gain & Losses at 89.8°			
25ft.H	3.5	3.75	4.0MHz
No loss	4.67	5.09	5.22dBi
w/ loss	4.66	5.08	5.12dBi
RF Loss	0.01	0.01	0.10dB
Coax loss - RG-213 Vf=0.66			

Figure 34

There were very small changes necessary in moving the capacitive coaxial Stubs from a low-end location on the Delta, up to the higher ends of Wires 4 and 5. These wires had to be extended, since the capacitive coaxial Stubs have less effect at this higher location. It would have been possible to leave these two wire lengths the same as the previous antenna and the two capacitive coaxial stubs lines L1 and L2 could have been extended, however this was not the arrangement chosen.

When these two capacitive coaxial stubs lines L1 and L2 are mounted on the radiator, as has been done in this antenna, it will be necessary to increase the length of Wires 1 and 3, since the larger diameter of the coaxial cable has less inductance, due to its diameter. Additionally the length of Wires 1 and 3 very likely will have to be increased in length slightly, to re-resonate the antenna and to center it in the bandpass.

This antenna has used two capacitive coaxial stubs in place of a Fixed Capacitor and has achieved substantially the same performance, low losses, as well as the same voltage ratings as the previous antenna and is to be considered a "High Rock Super-Delta Antenna".

75-80m Super-Delta Horizontal with Dual Stubs

The previous antenna utilized two capacitive coaxial stubs and they are effectively in series with each other and therefore the combined voltage rating is twice that of the basic RG-213 coaxial cable. Whenever these coaxial cables are placed in series, they must be twice as long, to achieve the same capacity. In the previous antennas that used the series arrangement for the stubs, their length was approximately 116" inches long. If a single stub can be used, its length would be cut in half, as shown in Figure 35. In this antenna, two stubs are used in a parallel configuration and the combined length of these two stubs are approximately 68" inches long in order to have the proper capacitive reactants.

Each individual Stub has a length of 34" inches and these are placed electrically in parallel, with each placed separately at the bottom of Wires 4 and 5. They are not in series however because of the crisscross wire connections of the two stubs, they are effectively in parallel and they are mounted on Wires 6 and 7.

When these stubs are inserted into the lower end of Wires 4 and 5, two additional Wires, numbers 6 and 7 were generated on which these 34" inch stubs are mounted. This was necessary since the coaxial braid diameter of the RG-213 coaxial cable is approximately 0.32 inches OD. With this larger diameter, the 34" inch lengths will have less inductance than a #14 AWG wire of the same length. The modeling program had to have Wires 6 and 7 specified at 0.32" inches, whereas all other wires are #14.

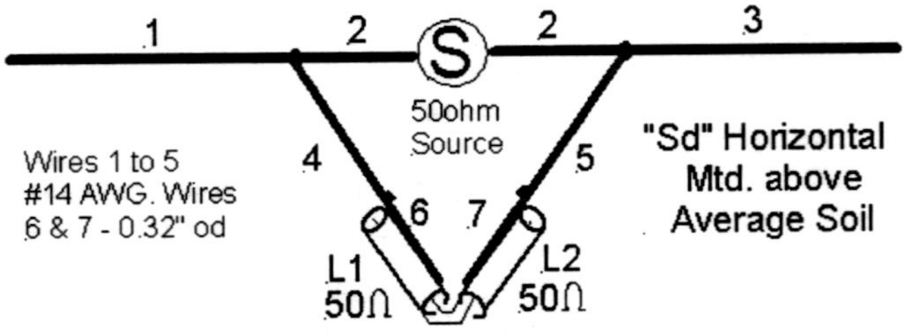

Figure 35

The Diamond configuration of Figure 35 shows this arrangement with the two 34" inch Stubs attached to Wires 6 and 7.

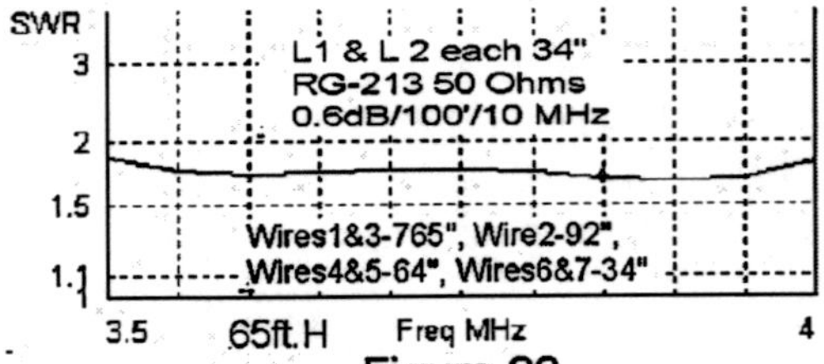

Figure 36

At a mounting height of 65' feet above Average-soil, the SWR plot of Figure 36 shows an SWR of approximately 1.8/1, which is flat across the entire 3.5 to 4MHz bandwidth and it is consistent with the previous antenna.

The Far-field RF radiation chart of Figure 37 shows good gains of 6.22, 6.18 and 6.07dBi with a major radiation lobe at 57.9°. Although the Short-line capacitive Stubs are not in a series configuration, their losses must be extremely low since this chart only shows coaxial losses of 0.02, 0.00 and 0.00dB across the 3.5 to 4MHz bandwidth.

Far-field Gain & Losses at 57.9°			
65ft.H	3.5	3.75	4.0MHz
No loss	6.22	6.18	6.07dBi
w/ loss	6.20	6.18	6.07dBi
RF Loss	0.02	0.00	0.00dB
Coax loss - RG-213 Vf=0.66			

Figure 37

Reducing the antenna height to a nominal 45' foot, as seen in Figure 38, the SWR of course has suffered slightly and is flat at an SWR of approximately 2.3/1, which is a satisfactory level for an antenna of this height.

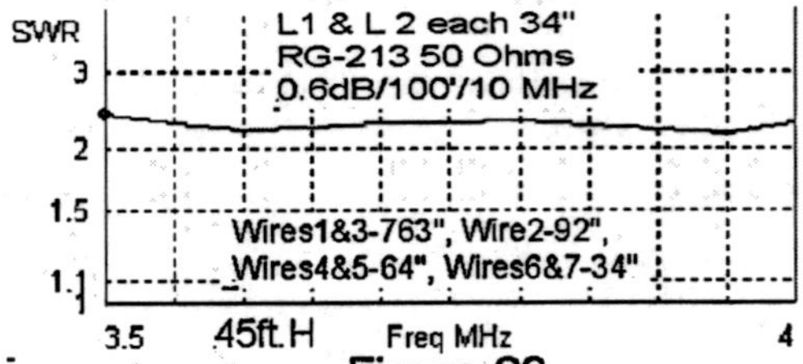

Figure 38

The Far-field gain and loss chart of Figure 39 shows slight increases in the RF gain at an angle of 89.9° and this increase is in-line with previous antennas of this type. At this height, the coaxial losses are once again very low at 0.02, 0.01 and 0.01dB across the entire bandwidth, making this a very efficient, well performing antenna at this height.

Far-field Gain & Losses at 89.9°			
45ft.H	3.5	3.75	4.0MHz
No loss	6.35	6.47	6.42dBi
w/ loss	6.35	6.46	6.41dBi
RF Loss	0.02	0.01	0.01dB
Coax loss - RG-213 Vf=0.66			

Figure 39

Reducing the antenna height further to 25' feet, as seen in Figure 40, has caused the SWR to deteriorate to a value of 3/1, which is in-line with the previous antennas in this chapter. Most Internal Antenna-Tuners are capable of improving this SWR to lower levels.

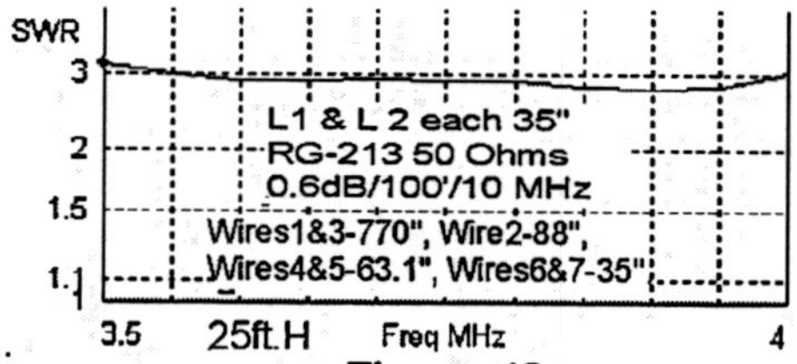

Figure 40

The Far-field RF radiation chart of Figure 41 shows that the RF gains have dropped to 4.52, 4.90 and 4.99dBi over the entire 3.5 to 4MHz bandwidth and the coaxial losses have dropped down substantially to 0.00dB.

Far-field Gain & Losses at 89.8°			
25ft.H	3.5	3.75	4.0MHz
No loss	4.52	4.90	5.00dBi
w/ loss	4.52	4.90	4.99dBi
RF Loss	0.00	0.00	0.01dB
Coax loss - RG-213 Vf=0.66			

Figure 41

Figure 42 is a chart providing information as to the level of RF voltages on the coaxial stubs L1 and L2, when using an RF transmitting power of 200 Watts. These voltages were measured when the antenna was mounted at a height of 45' feet and at frequencies of 3.5, 3.75 and 4MHz. These voltages range from 242 to 436 and to 570RMS across the bandwidth. The recommended coaxial cable RG-213 has a rated RF voltage of 3700 V and with this in mind, the voltage when using a power of 800 Watts will double, well within the RG-213 voltage specification of 3700V.

RF Vrms on L1&L2 at 200W Pwr			
45ft.H	3.5	3.75	4.0MHz
RF V	242	436	570Volts
L 1 and L 2, each 34" long			
Coax loss - RG-213 Vf=0.66			
Capacity per ft 30.8pF, at 3700 V			

Figure 42

In this antenna, the use of a Fixed Capacitor has been eliminated by the use of two Short-line Capacitive open-ended Stubs. In doing this, it allows trimming of the resonant frequency of the Diamond configuration, either by changing the length of both of the 34" inch lines L1 and L2 or in place of doing this, the length of Wires 4 and 5 can be extended or shortened slightly in length, so that the Diamond configuration is once again brought to the correct resonant frequency in order to achieve the best SWR. This "High Rock Super-Delta Horizontal Antenna" is one of the most adjustable, simplest and best performing antennas and above all, it is one of the most efficient. It is an antenna to be proud of.

160m Super-Delta Horizontal with Dual Stubs

The previous antenna, which covered the 3.5 to 4MHz band, was scaled to the 160m radio Amateur band which covers 1.8 to 2.0MHz and adjusted for the best SWR and is shown in Figure 43. During the modeling/scaling process, not only were the wires increased in length, but the Short-line Open ended Dual Stubs were increased from 34" inches, to lengths of 70" and 72" inches, depending on the radiator mounting height. Since RG-213, 50 Ohm coaxial cable is still being used; Wires 6 and 7 have been specified as the same 0.32" inches in diameter as used in the previous antenna.

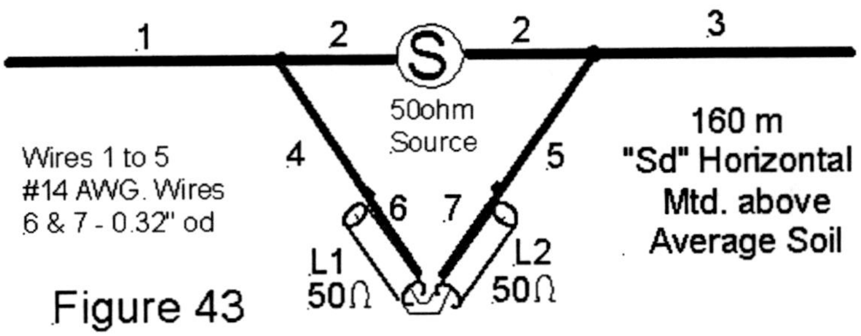

Figure 43

A basic Center Fed horizontal radiator was mounted at 80' feet and adjusted for the best SWR at a frequency of 1.9MHz. The SWR plot as shown in Figure 44 is provided, to give some sense as to what the SWR might possibly be when it was devoid of any Broad-Banding. The SWR was very low at 1.9MHz and deteriorated to well over 3/1 SWR or possibly 4/1. At the band edges it becomes problematic as to whether an Internal Antenna-Tuner will function and therefore Broad-Banding could provide better, more manageable SWR response.

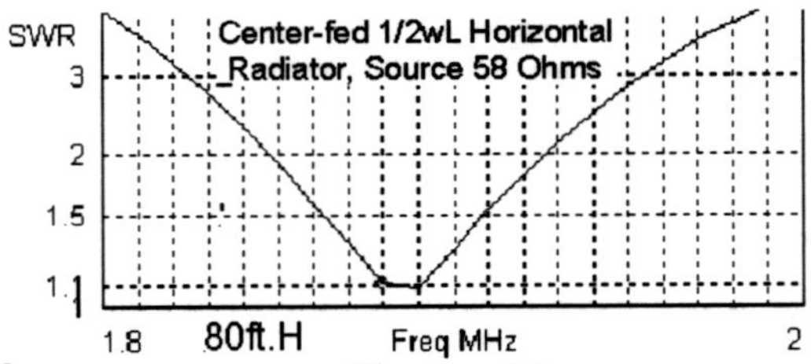

Figure 44

When this antenna is mounted at 80' feet and adjusted for the best SWR, the SWR plot of Figure 45 shows an SWR somewhat above 2/1 or approximately 2.3/1. This is the best that this antenna can achieve with the Broad-Banding provided. This antenna, which is mounted at 80' feet, is equivalent to a similar antenna in the 75-80m band mounted at a height of 40' feet and if you look back at the previous antenna at 45' feet, you will find that the SWR is approximately the same. Therefore, it is hard to expect a better SWR in the 160m band, when mounted at 80' feet. Nevertheless, this is a perfectly workable SWR for most uses in the 160m band.

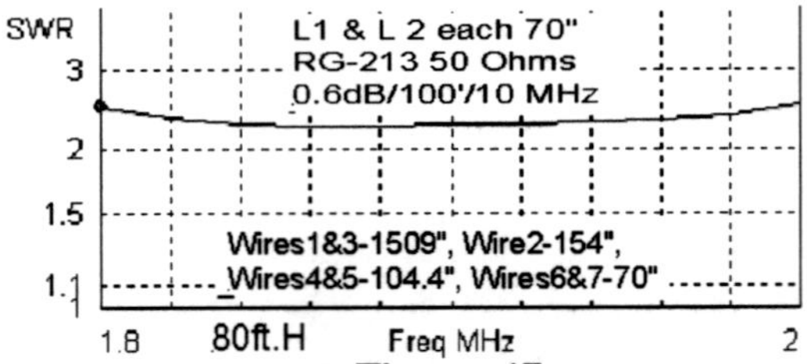

Figure 45

The Far-field RF radiation chart of Figure 46 shows that this antenna has excellent radiation, values of 6.64, 6.76 and 6.72dBi at a radiation angle of 57.9°. Even though the coaxial stubs are approximately twice as long, their losses are extremely low at 0.01, 0.00 and 0.01dB across the 160m bandwidth.

Far-field Gain & Losses at 57.9°			
80ft.H	1.8	1.9	2.0MHz
No loss	6.65	6.76	6.73dBi
w/ loss	6.64	6.76	6.72dBi
RF Loss	0.01	0.00	0.01dB
Coax loss - RG-213 Vf=0.66			

Figure 46

Lowering the antenna to 50' feet above Average-soil, the SWR has deteriorated to a Figure of 3/1, at which most Internal Antenna-Tuners will function properly to improve the SWR further. As stated previously, this is equivalent to having an antenna at one-half of this height in the 75-80m band. The previous antenna when mounted at 25' feet has a similar SWR. This new mode 160m antenna is shown in Figure 47 and it has a stub length of 72" inches for both lines L1 and L2.

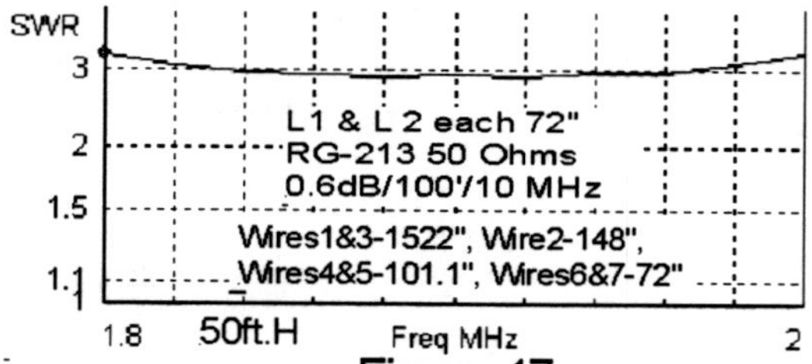

Figure 47

Far-field Gain & Losses at 89.7°			
50ft.H	1.8	1.9	2.0MHz
No loss	5.31	5.60	5.63dBi
w/ loss	5.31	5.630	5.626dBi
RF Loss	0.00	0.00	0.01dB
Coax loss - RG-213 Vf=0.66			

Figure 48

At a height of 50' feet, the Far-field RF gains have deteriorated, as expected, down to 5.31, 5.63 and 5.62dBi at a radiation angle of 89.7°. This is a relatively good SWR for this height and the coaxial losses are still almost nonexistent at 0.00, 0.00 and 0.01dB across the 160m bandwidth.

The RF voltages that are across each of the coaxial stubs L1 and L2 are shown in Figure 49. These RF voltages are 196, 534 and 660V RMS when a power level of 200 Watts is used and they will double at a power level of 800 Watts.

RF Vrms on L1&L2 at 200W Pwr			
50ft.H	1.8	1.9	2.0MHz
RF V	196	534	660Volts
L 1 and L 2, each 72" long			
Coax loss - RG-213 Vf=0.66			
Capacity per ft 30.8pF, at 3700 V			

Figure 49

This is another "High Rock Super-Delta Horizontal Antenna" that has all of the qualities such as simplicity, low cost of materials, easily adjusted, very high-efficiency and overall well performing. I suspect that this antenna may become a standard for use in the 160m band, since the two short stubs have a minimal of weight. Longer stubs are heavier and become a detriment after 160m because of the long antenna radiator length and will have a tendency to cause severe drooping. This antenna certainly can be called a "Super-Delta Horizontal Antenna".

160m Super-Delta Horizontal with a Fixed Capacitor

The most basic High Rock Super-Delta Horizontal Antenna, is one in which the configuration is brought to resonance by the use of a single Fixed Capacitor. This Capacitor has a particular tolerance and it is best to specify it to be very low and precise, so that the antenna wire lengths specified, along with this Capacitor, will probably resonate near the center of the selected bandwidth.

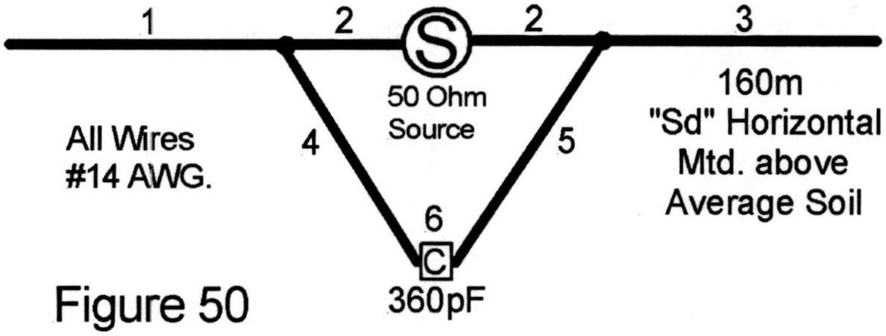

Figure 50

If the tolerance is inadequate, all is not lost and Wire 6, on which this Fixed Capacitor is mounted, can be raised or lowered, which causes Wires 4 and 5 to decrease or increase in length, in order to resonate the Diamond resonant circuit to the proper frequency. By using wire clamps on the ends of the Capacitor housing will allow these wires to slide through the clamp when adjusting the lengths of Wires 4 and 5, these clamps are then secured tightly.

Figure 50 shows such an antenna and it uses a Capacitor value of 360pF in order to resonate the Diamond resonant circuit and all of the wires have been increased in lengths to have this antenna operated on the 160m band.

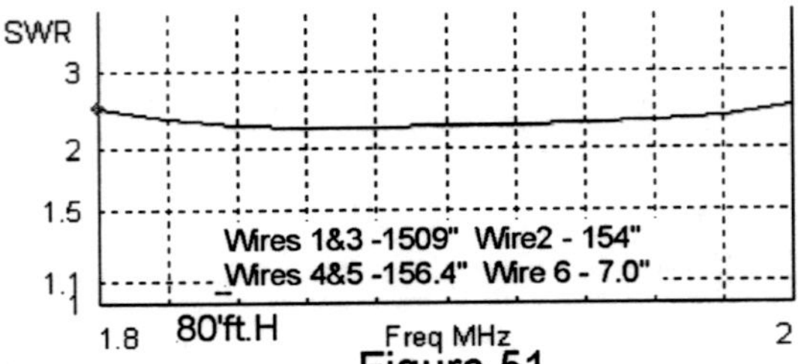

Figure 51

When this antenna is mounted at a height of 80' feet and the SWR has been adjusted to be symmetrical, the SWR of approximately 2.3/1 should be obtainable as shown in Figure 51. The previous antenna that uses coaxial stubs as a replacement for the Fixed Capacitors has substantially the same SWR value as seen in Figure 45.

The Far-field RF gain chart of Figure 52 shows gains of 6.65, 6.78 and 6.71dBi across the entire 160m bandwidth. These are extremely good Figures and since no coaxial cables are used, there is nothing in this antenna to create loss, except the Capacitor and the antenna wires themselves.

Far-field Gain & Losses at 57.9°			
80'ft.H	1.8	1.9	2.0MHz
RF gain	6.65	6.78	6.71dBi

Figure 52

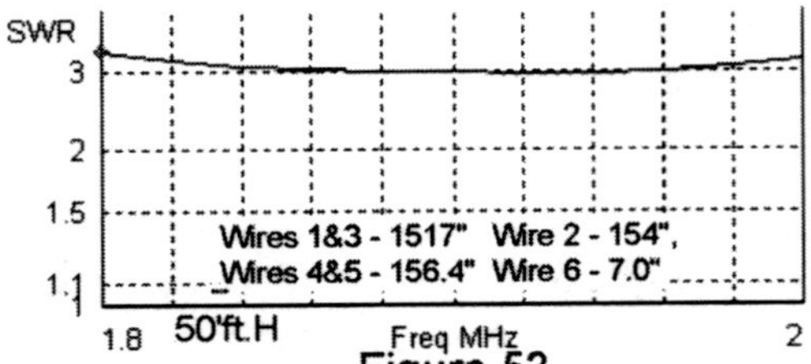

Figure 53

When this antenna is lowered further to a height of 50' feet, the SWR will deteriorate to an SWR of approximately 3/1, as shown in Figure 53. This is substantially the same as the previous antenna at the same height.

The RF gains of the chart shown in Figure 54 have gains of 5.31, 5.61 and 5.64dBi at an angle of 89.7°. Considering that this antenna height of 50' feet in the 160m band is equivalent to 25' feet in height in the 75-80m band, therefore these Far-field RF gains are very good.

Far-field Gain & Losses at 89.7°			
50ft.H	1.8	1.9	2.0MHz
RF gain	5.31	5.61	5.64dBi

Figure 54

This "High Rock Super-Delta Horizontal Antenna" is an excellent performer and it has one of the lightest physical loads. It contains a single Capacitor, rather than two lengths of RG-213 coaxial cable which would have 140 or 144" inches of length. You can choose either version and have excellent performance and very high efficiency.

This is the last of the Horizontal antennas to be presented, Chapter 4 will devote presenting "High Rock Super-Diamond Inverted V Antennas", which also have excellent performance and high-efficiency.

Super Max-Dipole Antennas

By Edward J. Shortridge W4JOQ

Chapter 4

Super-Diamond Inverted-V Antennas

Introduction

This chapter will present full size Super Max-Diamond Antennas that directly provide a 50 Ohm Source, whereas, in previous chapters many of the Diamond configurations were reduced in physical size and provided Source impedance well below 50 Ohms. Although these antennas have certain advantages, most all of them required some form of matching in order to provide the proper performance. These latest Diamond configurations are considerably larger in physical size in order to provide this 50 Ohm Source.

All of the Diamond antennas have a configuration that is quite versatile and are capable of having separate means of adjusting the common coupling wire lengths to achieve the best SWR response and separate wires that can be adjusted, along with a resonating Capacitor value or capacitive Stubs in order to resonate the total Diamond configuration to the proper frequency, near the center of the desired bandwidth.

These antennas are easy to construct, adjust and can provide low SWR bandwidths quite easily and it is expected that they will become the antenna of choice, when considering using Inverted V Radiators. These antennas are most suited to the 75-80m and 160m Amateur radio bands, where wire antennas are used and therefore all of the antennas in this chapter will cover either the 75-80m or the 160m bands.

These Diamond configurations will be shown in a normal-view, with the Source located at the Apex of the Inverted V. Then the Diamond will be rotated, such that the Source will be at the bottom of the Diamond. The most basic configuration will use a single Fixed Capacitor to resonate the diamond. Open-ended Capacitive Stubs will be used as a substitute for the Fixed Capacitor. These stubs will be shown mounted in a variety of locations on the Diamond, permitting great flexibility of the selection in which you can choose the variation that you consider most adaptable and applicable to your needs.

Throughout this chapter Far-field RF radiation charts, which shows the RF gains, with and without the coaxial cable losses, the difference in these losses will be listed in terms of decibels and these will be shown at different antenna heights, giving you an excellent idea of how efficient these Diamond antennas can be.

The Need for Low SWR Antennas

Whenever you have a high SWR, as measured at the antenna Feed-Point and you use a lengthy 50 Ohm transmission line, you will have SWR coaxial losses that can never be removed by an Antenna-Tuner located far from the antenna feed-point, such as in the Radio room. Where the Antenna-Tuner is a separate device or built-in to a transceiver, it will make no difference. The most that the Antenna-Tuner can do is to match the impedance present at the coaxial cable and to the transmitter desired impedance of 50 Ohms, but this will in no way reduce the SWR and attendant increased losses on the transmission line.

The only means of overcoming this problem and reducing these transmission line losses to negligible values would be to use a remotely-controlled Antenna-Tuner which is located near the antenna itself.

If you were to design an antenna that is Broad-Banded and it uses long Broad-Banding lines that are 1/4 wavelength long, it is possible to have an antenna with an extremely good SWR but because of the long resonant 1/4 wavelength line, you might have coaxial losses in this line that could be as high as 1.6 to 2dB and not be aware of such losses. Once again, no Antenna-Tuner can tune out these losses, whether the tuner is in the Radio Room or a remote position at the antenna. With 2dB of loss, over one quarter of your total power is being dissipated in small portions of the coaxial cable, primarily at that high current and high-voltage ends.

Therefore, it is extremely important to be able to measure such losses as the antenna is being developed. One-quarter wavelength lines generally have high voltages at the open or high impedance end and higher currents at the shorted end, both of which contribute greatly to coaxial losses.

The Diamond Antenna configuration either uses a Fixed Capacitor to resonate the Diamond configuration or instead uses a Short-length coaxial Stub that is only long enough to supply the same capacitive reactants as that of the Fixed Capacitor. This is not the long resonant line and the voltages on this line are very moderate, resulting in very low coaxial losses. Additionally, some of the Diamond Antennas use a Fixed Capacitor and if a good grade of Mica Capacitor is used it can have a Q value approaching 5000, which is almost totally lossless.

Super-Diamond Antennas for the 75-80m and the 160m bands are being presented that include all of the possible variations, such as ones using a discrete Fixed Capacitor or Short-line Open-ended coaxial Stubs, which are located at different positions on the Diamond configuration.

75-80m Normal Diamond Antenna with a Fixed Capacitor

This is an Inverted V antenna with a 35° radiator slope and it has the normal 50 Ohm Source located at the very Apex of the configuration, which uses a Diamond shaped loop, which consists of Wires 2, 3, 5 and 6 as shown in Figure 1. This Diamond loop is brought to resonance near the center of the desired bandwidth by using a 220pF Fixed Capacitor, placed at the bottom junction of Wires 5 and 6. The tolerance of this Capacitor should be 5% or less, so that the lengths of Wires 4 and 5, which are lengthened and or shortened to compensate for this tolerance, do not have to have their lengths changed excessively. #14 AWG wire is used throughout this antenna.

In order to facilitate the shortening or lengthening of Wires 5 and 6 the Fixed Capacitor housing should have wire clamps attached to it, so that the wires can readily slide through the clamp and when the desired wire lengths are achieved, the clamps can be tightened securely.

When this antenna is mounted and the 35° sloping radiator ends are pulled tightly, the Diamond loop will expand and to assure that the desired shape is maintained, an insulated restraining cord should be placed between the lower ends of Wires 2 and 3, so that the loop cannot be expanded further. Dimension "D" shows the location on Figure 1 and the actual dimensions will be indicated on each SWR plot that will be shown. When the normal figure configuration is used, in which the Source is located at the Apex of the Inverted V, the resonating Capacitor is then located at the lower end of the Diamond and this presents a problem.

Since there is no transmission line attached to the Capacitor housing, it is free to move and change the shape of the Diamond configuration as well as its frequency and therefore this requires some correction.

It is expected that a 50/50 Balun will be used to create a balanced feed at the Source and when the 50 Ohm coaxial transmission line is attached, it can be brought down and have the Capacitor housing attached to it physically, so that this housing is tied securely and cannot readily move its position.

Other Diamond configurations that use Short-line Open-ended coaxial stubs have a similar problem that is even greater. Certain versions will not have any sort of housing at the bottom end of Wires 5 or 6 and an insulated support from this point of the Diamond must also be attached to the downward transmission line in order to stabilize the bottom end of the Diamond. It is expected that the lower ends of Wires 4 and 5 will also utilize a wire clamp to facilitate lengthening and shortening the wire lengths and have simple sliding motion and then be tightened to secure the connection.

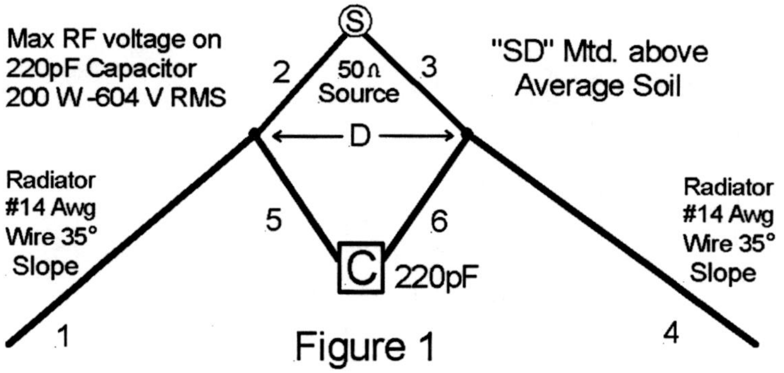

Figure 1

When this antenna is mounted at a height of 85' feet above Average-soil and the Diamond loop adjusted properly by altering the lengths of Wires 5 and 6, as necessary to achieve the best SWR plot, is as shown in Figure 2. This particular plot has an SWR slightly above 2/1, which is slightly above a more typical value closer to 2/1.This particular antenna has a dimension "D" length of 64.3" inches and the lengths of all of the other wires that comprise this antenna are shown in Figure 2.

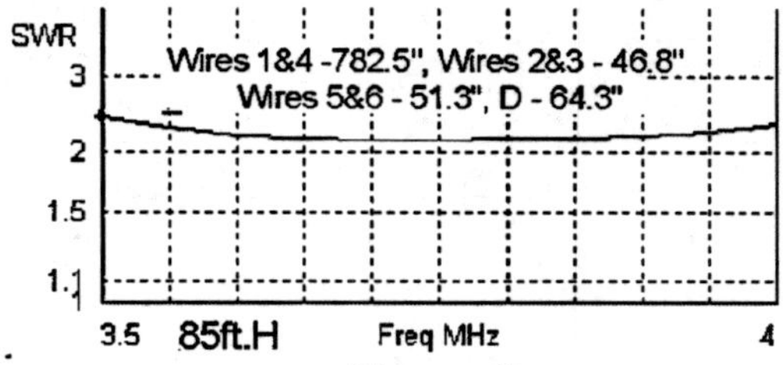

Figure 2

Far-field Gain & Losses at 55.9°			
MHz	3.5	3.75	4.0MHz
RF gain	5.29	5.28	5.28dBi

Figure 3

Since a Fixed Capacitor of 220pF is used to resonate the Diamond, it will have substantially no loss compared to other antennas that utilize some length of coaxial cable which might be used as a substitute. Therefore no RF losses are specified in the chart of Figure 3, which indicates RF gains of 5.29, 5.28 and 5.28dBi across the 3.5 to 4MHz band and the maximum radiation is at an angle of 55.9° above the horizon. All of these are good indications of a very well performing antenna.

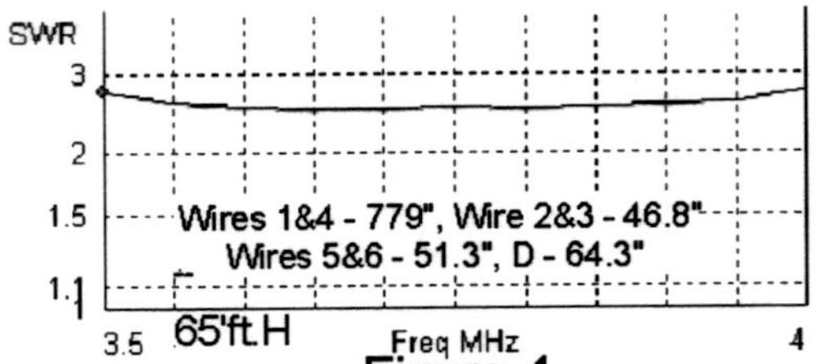

Figure 4

Lowering the antenna to 65' feet causes the SWR to deteriorate and be symmetrical. Due to this lower height, the capacity to ground has increased and the radiator ends must be shortened, as indicated on the SWR plot, in order to re-center the SWR and have it in a symmetrical shape as shown in Figure 4.

This antenna was lowered as much as possible and still keeping the radiator ends up at a reasonable height of 7.6' feet above ground, which is well out of reach. This places the apex of the antenna at a height of 47' feet above ground, which is the lowest height recommended when using a radiator slope of 35°. The resulting antenna has its radiator ends shortened once again in order to use this height.

Figure 5 shows an SWR slightly lower than 3/1, which is the best that generally can be obtained with most Diamond antennas mounted at this height and having a 35° radiator slope.

At a height of 47' feet, the Far-field RF gain has decreased to values of 3.41, 3.67 and 3.83dBi, with the maximum radiation at an angle of 89.7°, as shown in Figure 6. These figures are quite typical of most similar antennas at this mounting height.

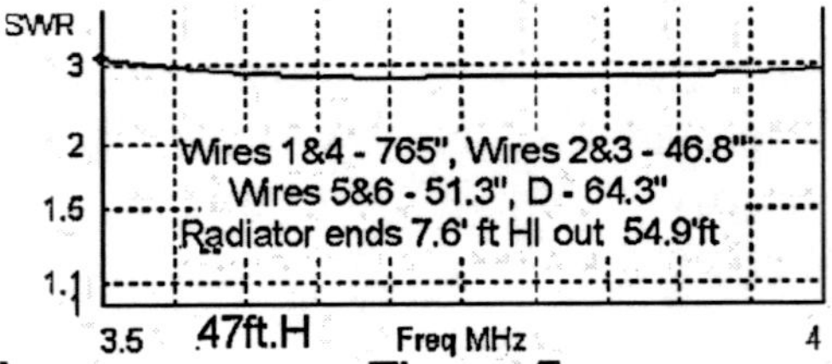

Figure 5

Far-field Gain & Losses at 89.7°			
MHz	3.50	3.75	4.0MHz
RF gain	3.41	3.67	3.83dBi

Figure 6

In order to determine the voltage stresses across the 220pF resonating Capacitor, voltages were measured at three frequencies across the 75-80m band, as shown in Figure 7, which are 166, 394 and 604V rms, which were measured and calculated for transmitter output power of 200 Watts. At a power level of 800 Watts, these voltages will double. These voltages are listed as RMS and the peak voltage will be approximately 1.414 times this value.

RF volts on capacitorwith 200 W			
MHz	3.5	3.75	4.0MHz
RF volts	166	394	604rms

Figure 7

When using Inverted V radiators, the SWR bandwidth will be constricted, as shown in this antenna when using the "Critically Coupled Broad-Banding Method" and it is unusual to obtain values that differ from those shown, except in a very few unusual cases. In general creating a radiator slope of 45°, causes even greater difficulties in obtaining the same SWR curve.

Suggestions for Trimming Diamond Antennas

When you wish to construct and mount a Diamond antenna, you may encounter a situation in which the SWR differs somewhat from the intended symmetrical SWR. This may occur due to the accuracy of each of the wires that comprise the antenna or it might just be the tolerance of the Fixed Capacitor or the length of a coaxial stub. In order to provide information on how each of the wires can affect the final SWR obtained, the list showing how each of these wires contributes to the symmetry of the SWR bandpass will be provided.

The first Super-Diamond Antenna shown in this chapter is in Figures 1 and 2 which were selected to demonstrate how the trimming of certain wires will have on the final SWR. A copy of the SWR plot of Figure 2 is once again shown mounted at a height of 85' feet above ground and is titled Figure 7a.

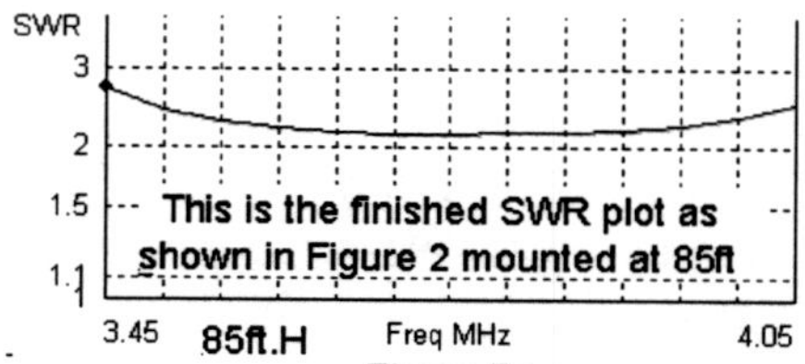

Figure 7a

In order to show how much the SWR is shifted a figure of 8" inches was selected to trim the lower ends of Wires 5 and 6. These two wires were shortened by 8" inch lengths and the SWR plot of Figure 7b is presented to show the results of this 8" inch amount of detuning. The antenna was scanned an extra 50 kHz below and above the 75-80m band in order to see where the SWR deteriorates at a faster rate.

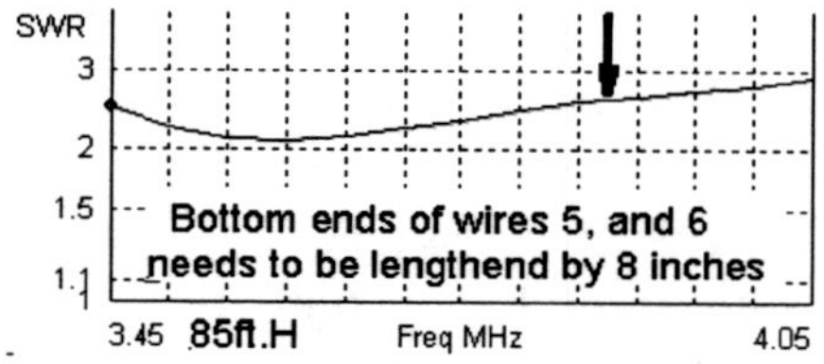

Figure 7b

Due to these change in lengths, the upper end of the SWR bandwidth increased, approaching an SWR of 3/1. It is now known how much the SWR will change with this shortening of Wires 5 and 6. If these two wires are once again increased in length by 8" inches, which would be necessary to trim the antenna, the original symmetrical SWR bandwidth can once again be obtained. The down-pointing arrow on Figure 7b shows the direction in which the SWR must be improved.

Figure 7c shows the reverse condition, in which the SWR is altered by having the lengths of Wires 5 and 6 increased by 8" inches in length. This change has caused the SWR to improve at the upper end of the band and the low end of the plot has deteriorated to an SWR of 3/1. This is an unacceptable condition, so the antenna must be trimmed, shortening the bottom ends of Wires 5 and 6, to the original length by 8" inches, as shown in Figure 7a.

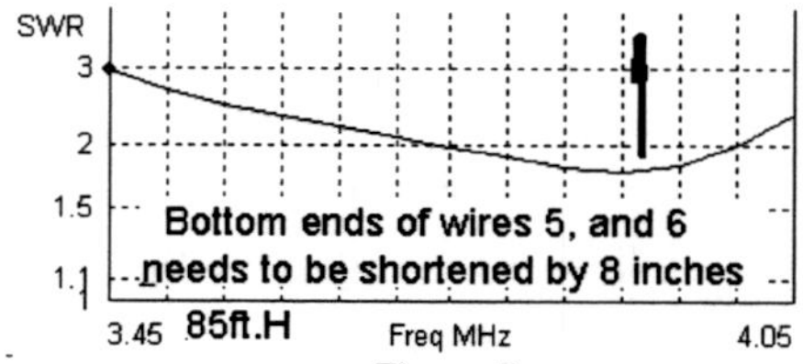

Figure 7c

Figure 7d once again uses the symmetrical SWR of Figure 7a and the ends of the radiator Wires 1 and 4 are shortened by 8" inches. This shifts the antenna resonance to a higher frequency, which improves the SWR at the upper end of the bandwidth, whereas the lower end of the bandwidth deteriorates to an SWR of 3/1. If you have this SWR condition you must increase the lengths of the radiator Wires 1 and 4 by an appropriate amount and in this particular antenna this length will amount to 8" inches. The bold arrow shows the direction of the required shift in frequency.

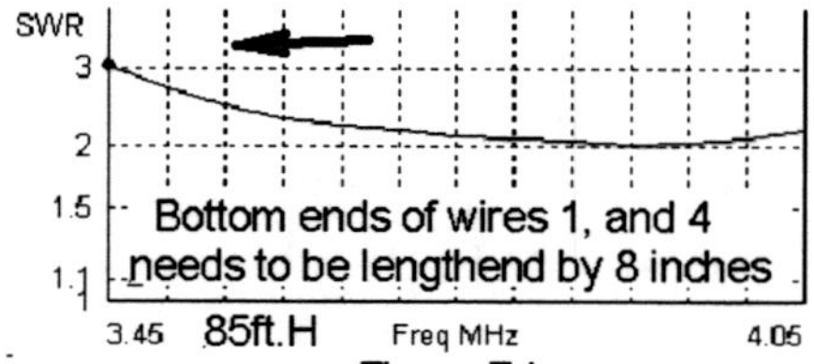

Figure 7d

Figure 7e shows the exact reverse, whereas the lengths of the radiator ends are increased by 8" inches, thus causing the SWR at the low end of the bandwidth to improve and this causes the SWR at the high end of the bandwidth to deteriorate to an SWR of approximately 3/1. Since we have increased the lengths by 8" inches, we automatically know that if we recognize this condition, we need to reduce the radiator ends by the same 8" inches and the bold arrow indicates the direction in frequency that is required.

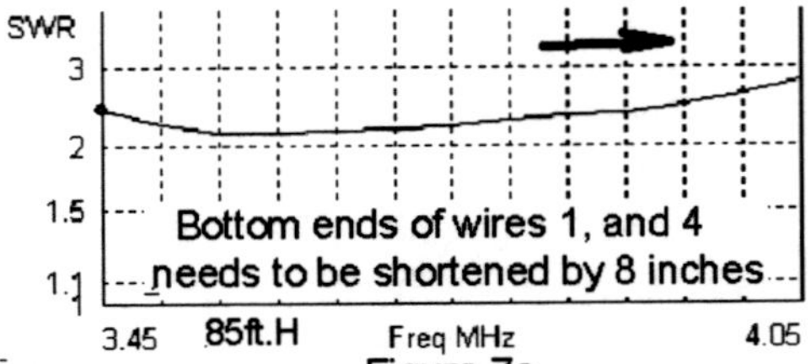

Figure 7e

This antenna utilizes a 220pF Capacitor which is used to resonate the Diamond configuration to a point near the center of the intended bandwidth. Unfortunately, all Fixed Capacitors have tolerance and we have recommended a 5% tolerance, which amounts to plus or minus 11pF. If the actual Capacitor that you use is at the extreme end of this tolerance, which could be 209pF or 231pF, you can expect some detuning of the Diamond loop and two examples will be presented to show what the SWR would look like when using Capacitors at the ends of this tolerance.

Figure 7f shows the SWR plot of an antenna in which the 220pF resonating Capacitor was changed to a value of 231pF, which is a + 5% value. This causes the upper end of the SWR response to improve, while the lower end has deteriorated. To correct this condition, a closer tolerance Capacitor can be used, or the bottom ends of Wires 5 and 6 can be reduced in length by an amount necessary to bring the Diamond loop back to a resonance at the band center. The bold arrow shows the direction that the SWR should be moving to correct this situation.

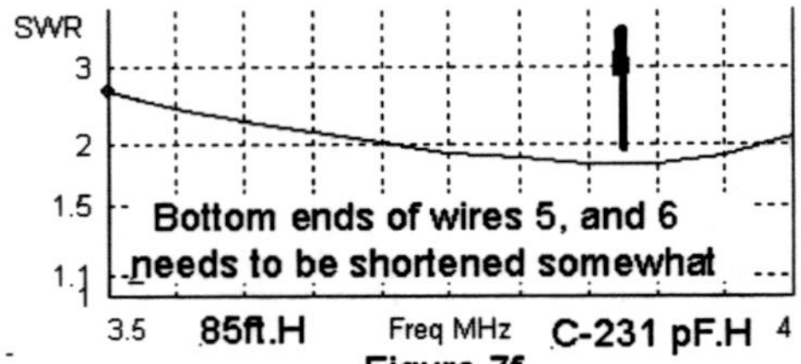

Figure 7f

Figure 7g shows a Diamond antenna in which the Fixed Capacitor has been altered to be at a value of 209pF, which is at the end of the -5% value. This has caused the SWR at the high end of the bandwidth to deteriorate to approximately 3/1. To correct this, a closer tolerance Capacitor can be used or the bottom lengths of Wires 5 and 6 can be extended by a sufficient amount to compensate for this Capacitor low end tolerance.

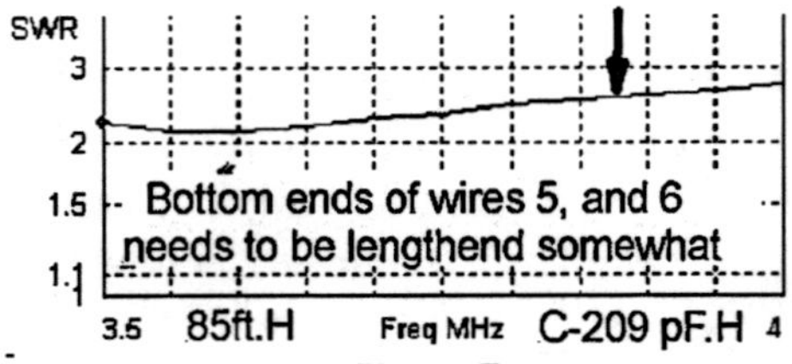

Figure 7g

Wires 2 and 3 control the amount of coupling between the resonant Diamond configuration and the radiator. The lengths chosen are values that give the best SWR over the entire 75-80m bandwidth. If the lengths of these two wires are shortened in order to reduce the amount of coupling, it will have a tendency to improve the SWR in the band center and have it deteriorate somewhat at the band ends. By shortening these two wires, the symmetry of the SWR bandwidth will no longer be flat. Increasing the length of Wires 5 and 6 will be necessary to once again bring the Diamond loop back into resonance at the band center, in order to once again obtain asymmetrical bandpass.

75-80m Rotated Diamond Antenna with a Fixed Capacitor

The first antenna shown in this chapter was a Normal Diamond Configuration, as shown in Figure 1 and the Diamond Loop, which consists of Wires 2, 3, 5 and 6 which are rotated 180° on an axis located at the lower ends of Wires 2 and 3. This in effect places the Source at the bottom end of the Diamond and the 220pF Capacitor is then located at the Apex of the Diamond, which is the highest point on the antenna as shown in Figure 8.

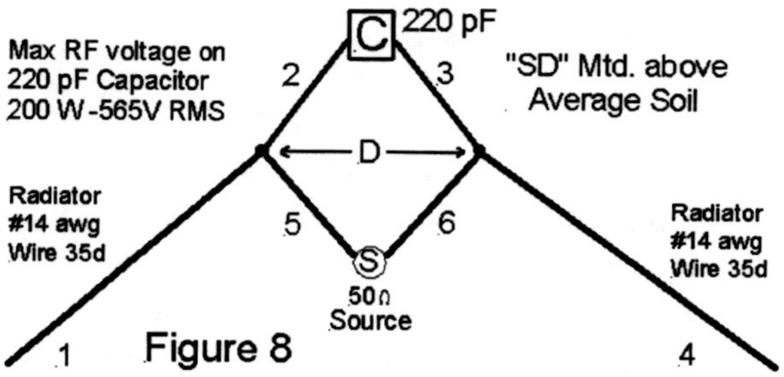

Figure 8

After rotating the Diamond and readjusting the antenna to 85' feet above ground, it was found that the antenna Diamond loop was under coupled from the radiator, producing an SWR that was low in the center portion of the band and it had much higher SWR at the band edges. Therefore; Wires 2 and 3 had to be altered from 46.8" inches to 49.8" inches and Wires 5 and 6 had to be altered from 51.3" inches to a length of 52.1" inches in order to obtain the very best SWR bandwidth. The resulting SWR is very close to 2/1 across most of the bandwidth, as shown in Figure 9, which is somewhat better than the previous antenna shown in Figure 1.

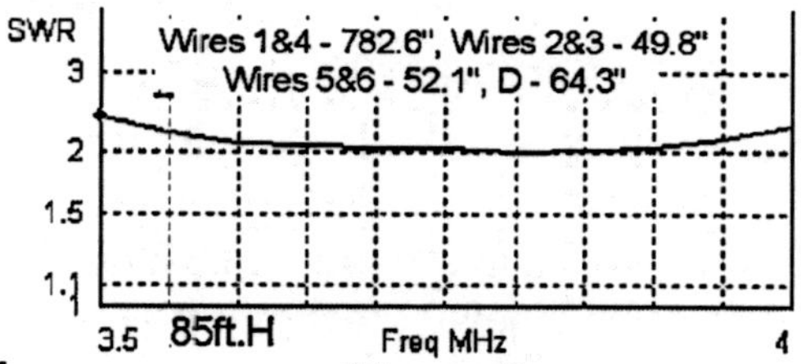

Figure 9

Earlier in this book the same phenomena was noticed and it appeared to be caused by a change in the coupling. It appeared that in the first case, which is the normal configuration, there was added Mutual Inductive coupling between the Diamond loop and the radiator and when the configuration was rotated, this coupling was greatly reduced and the two resonant circuits which are the radiator and the Diamond loop where then slightly under coupled.

After the antenna has been rotated, as shown in Figure 8, it was necessary to increase the common coupling Wires 2 and 3 from 46.8" to 49.8" inches, thereby increasing the coupling between the Diamond loop and the radiator. This change causes the loop to be detuned and to correct this; Wires 5 and 6 had to be decreased from 51.3" inches to 52.1" inches. The change for Wires 5 and 6 were a reverse from what was expected to be needed.

The Far-field RF gains as shown in Figure 10 shows gains of 5.18, 5.20 and 5.08dBi across the entire bandwidth from 3.5 to 4MHz. These gains are slightly lower than those shown in Figure 3, which are for the Normal configuration. This can be expected, since the radiator center portion consisting of Wires 2 and 3 have been decreased in height when the Diamond is rotated. This reduction in gain is quite small and insignificant.

Far-field Gain & Losses at 56.7°			
MHz	3.5	3.75	4.0MHz
RF gain	5.18	5.20	5.08dBi

Figure 10

When this antenna is lowered to a height of 65' feet, the lengths of the radiator end Wires 1 and 4 had to be changed in length from 782.6" inches to 777.6" inches to achieve asymmetrical bandpass. The SWR averages approximately 2.5/1 across the entire bandwidth, as shown in Figure 11. This is a very acceptable SWR compared with other antennas.

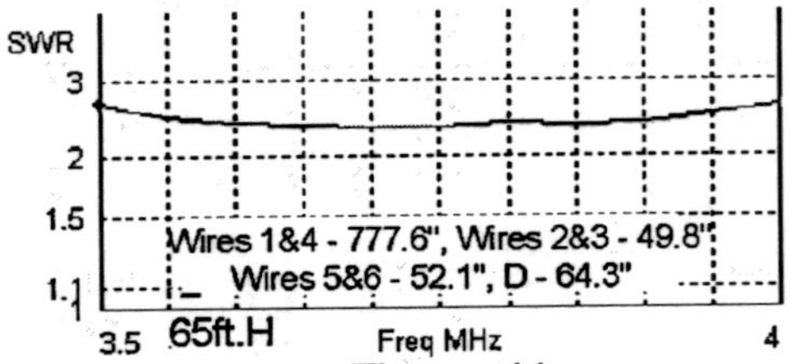

Figure 11

This antenna was lowered as much as possible, to the point where the radiator ends are 6.8' feet high and stretching out to a length of 55.1' feet, as shown in Figure 12. The 220pF Capacitor, which is located at the apex of the antenna, is at a height of 47' feet and after the radiator ends were shortened to a length of 768.6" inches, the SWR is well below 3/1 across the entire bandwidth, as shown in Figure 12.The RF gains, as shown in Figure 13, are 3.16, 3.45 and 3.50dBi, which are very good and typical of other Diamond antennas mounted at approximately the same height.

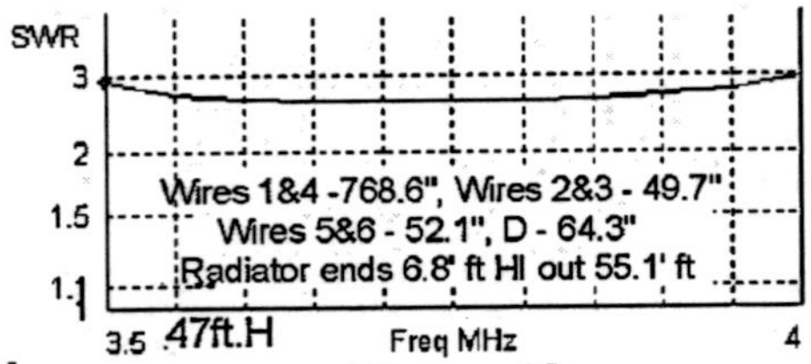

Figure 12

Far-field Gain & Losses at 89.7°			
MHz	3.50	3.75	4.0MHz
RF gain	3.16	3.45	3.50dBi

Figure 13

The voltages across the 220pF Capacitor must be known in order to determine whether the voltage rating of this Capacitor has been selected correctly. The chart of Figure 14 indicates if an RF power level of 200 Watts is used, the voltage appearing across this Capacitor will vary from 165 Volts, 473V and 565V across the 3.5 to 4MHz bandwidth. If the RF power levels are increased to 800 Watts, these voltages will double.

RF volts on capacitorwith 200 W			
MHz	3.50	3.75	4.0MHz
RF volts	165	473	565rms

Figure 14

Having the Diamond loop rotated puts the Capacitor housing at the Apex of the Inverted V and the transmission line is now capable of being attached to the Source which is now at the lower end of the loop. Having the transmission line attached assures that there is sufficient tension to prevent the lower end of the loop from moving and distorting the loop configuration but the radiator ends must be pulled tight to the point where the "D" insulated restraining cord is pulled tight.

75-80m Normal Diamond Antenna with Low Short-line Stubs

The first antenna shown in this chapter was a Normal configuration which used a 220pF Capacitor to resonate the Diamond loop; it is possible to substitute Short-line Open-ended coaxial stubs in place of this Capacitor, as shown in Figure 15. A single coaxial cable cannot be used since it will cause the Diamond loop and the entire antenna to be unbalanced. When this Short-line stub is to be used, a second identical Stub must be used with it to form a balanced twin stub. These two Stubs are in series, so the length of each stub must be twice as long as would be the case if a single Stub was to be used. 50 Ohm RG-213 coaxial cable should be used for the stubs and it has a specified loss of 0.6dB/100'ft/10MHz.

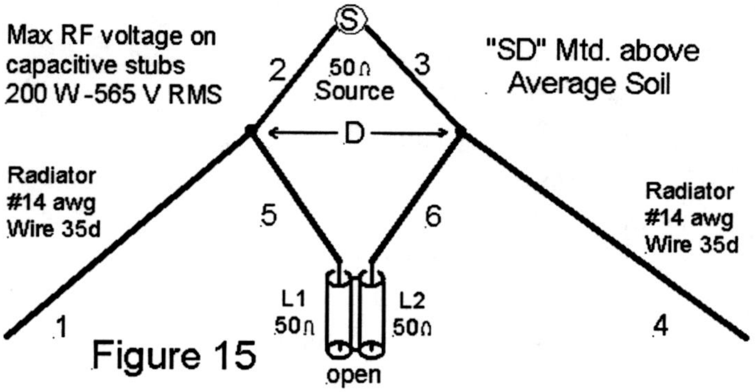

Figure 15

It is expected that a 50/50 Balun will be used at the Source, which is located at the apex of the Diamond loop. When a 50 Ohm transmission line is connected to the Balun, it can run downward and have the twin stubs attached securely to it, readily securing it in place.

When this antenna is mounted at a height of 85' feet, it will have an SWR almost identical to the one in which a 220pF Capacitor was used and this plot is shown in Figure 16. Each of the Short-line, Open-ended coaxial Stubs, lines L1 and L2 are 158" inches of 50 Ohm RG-213 coaxial cable, which has losses of 0.6dB/100'ft/10MHz.

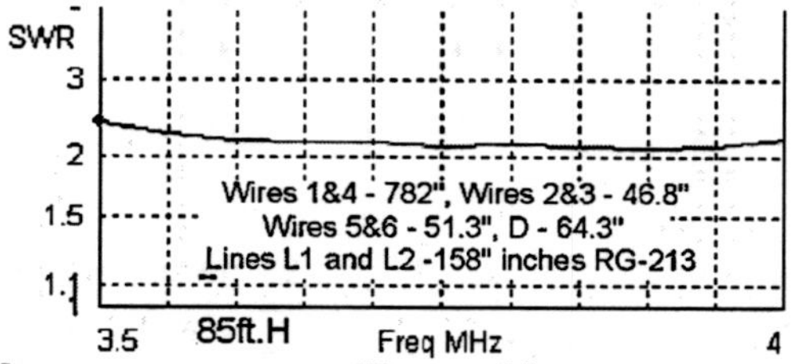

Figure 16

Far-field Gain & Losses at 55.7°			
85'ft.H	3.50	3.75	4.0MHz
no loss	5.26	5.28	5.28dBi
w/ loss	5.25	5.20	5.20dBi
RF Loss	0.01	0.08	0.086dB
Coax-0.6dB/100'/10MHz Vf=0.66			

Figure 17

The Far-field RF gain and loss chart of Figure 17 shows gains of 5.25, 5.20 and 5.20dBi across the entire 3.5 to 4MHz bandwidth and the coaxial stub losses are 0.01, 0.08 and 0.086dB. All of this shows that this antenna can perform well and has extremely low coaxial losses, making it an extremely efficient antenna.

Lowering the antenna to 65' feet, as shown in Figure 17a, the SWR deteriorated to approximately 2.6/1, which is a perfectly acceptable SWR for this height. Of course, each time that the radiator is lowered in height, the capacity to ground increases, requiring that the radiator ends are shortened, as was the case in this particular antenna.

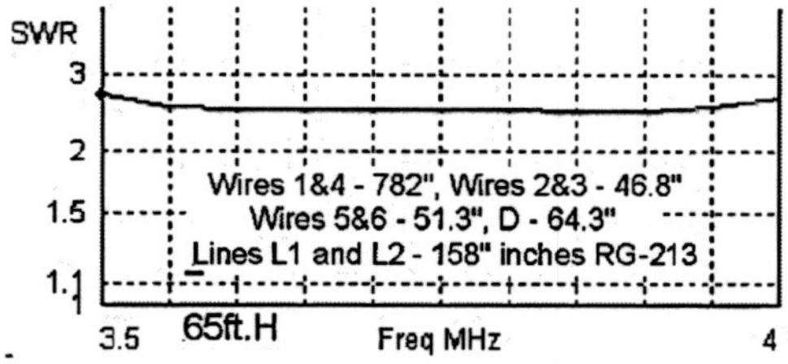

Figure 17a

This antenna was lowered as much as possible and at a height of 47' feet, as shown in Figure 18, the radiator ends extended down to a height of 7.4' feet and extend outward 55.25' feet. The SWR deteriorated slightly under an SWR of 3/1, which is adequate for Internal Antenna-Tuners to be capable of operating and improving the SWR.

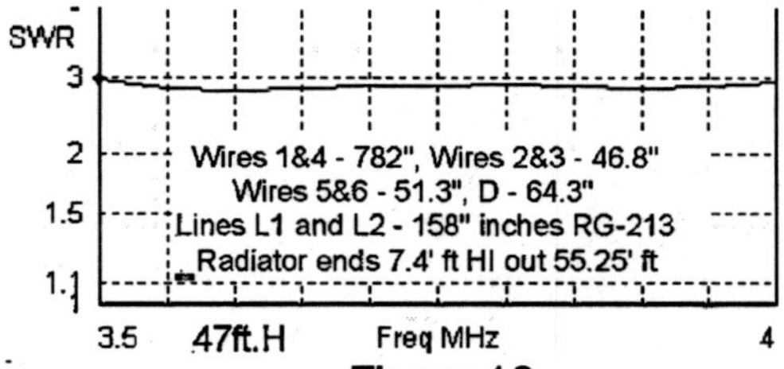

Figure 18

The Far-field RF gain and loss chart of Figure 19 shows gains of 3.37, 3.64 and 3.68dBi at a radiation angle of 89.9° above horizontal. These are good gain figures and the losses are 0.01, 0.02 and 0.11dB, which is very low.

Far-field Gain & Losses at 89.9°			
85ft.H	3.5	3.75	4.0MHz
no loss	3.38	3.66	3.79dBi
w/ loss	3.37	3.64	3.68dBi
RF Loss	0.01	0.02	0.116dB
Coax-0.6dB/100'/10MHz Vf=0.66			

Figure 19

This antenna clearly demonstrates that Short-line Open-ended coaxial stubs can be used as a substitute for the 220pF Capacitor as previously used. This substitute has not affected the SWR in anyway and the losses are quite low allowing this antenna to still be classified as a highly efficient antenna and can be called a Super-Diamond Antenna.

75-80m Normal Diamond Antenna with Hi Stubs

Another method of placing Stubs on the Diamond loop is the Hi Stubs configuration, in which the stubs are attached to the radiator, as shown in Figure 20. In order to model this configuration 2 extra Wires, 7 and 8 are placed on the radiator and the wire diameters for those two are specified as 0.32" inches, which is the diameter of the coaxial braid. The lengths of these two lines must be specified to be exactly the same length as each of the Short-line Open-ended coaxial stubs. When building this configuration, Wires 7 and 8 will actually be #14 AWG wire and the Stubs will be attached and supported by this wire diameter.

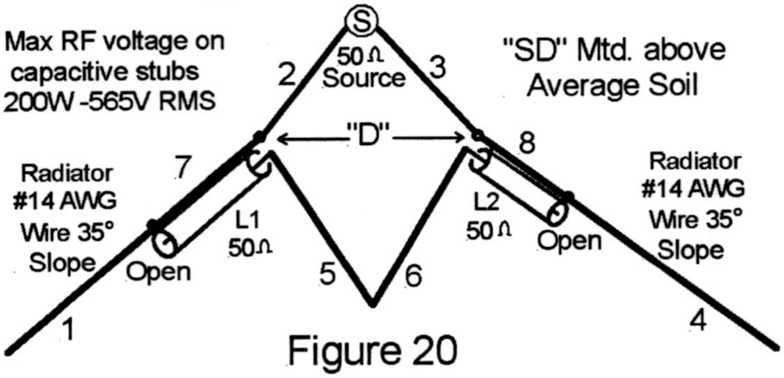

Figure 20

The lower ends of Wires 5 and 6 do not contain housing, nor connect the wires to a transmission line and the wires can easily be swayed and moved by the wind and they must be secured in some manner. A recommended 50/50 Balun should be used at the Source and when the 50 Ohm transmission line is attached, the lower ends of Wires 5 and 6 should be secured to the downward transmission line, making certain that these wires are well insulated from the transmission line. The 2 Stubs should have their coaxial braids attached from both ends to the radiator Wires 7 and 8.

At an antenna height of 85' feet, this antenna has an SWR slightly higher than 2/1, which is in-line with the previous Diamond antennas as was shown so far. This plot as shown in Figure 21 is perfectly flat and with a little extra modeling, might be capable of some slight improvements particularly across the main center portion of the bandwidth.

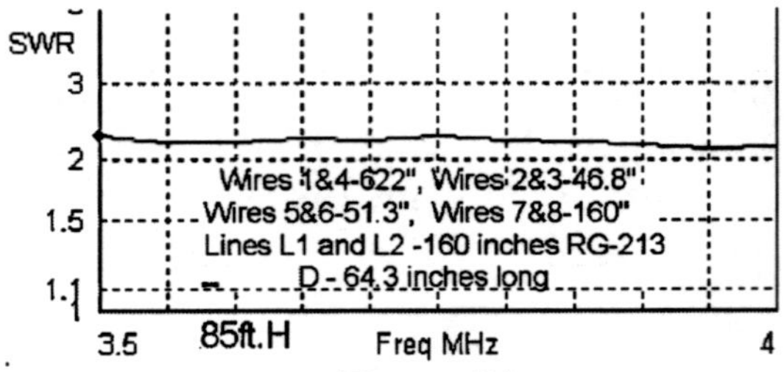

Figure 21

Far-field Gain & Losses at 55.7°			
MHz	3.50	3.75	4.0MHz
no loss	5.30	5.30	5.29dBi
w/ loss	5.28	5.28	5.22dBi
RF Loss	0.02	0.02	0.076dB
Coax-0.6dB/100'/10MHz Vf=0.66			

Figure 22

The Far-field RF gain and loss chart of Figure 22 has very good RF gains of 5.28, 5.28 and 5.22dBi at a radiation angle of 55.7° and once again the losses are quite low at 0.02, 0.02 and 0.07dB. Having the Stubs mounted on the radiator has not in any way had any adverse effects on the antenna and these particular stubs are 160" inches in length, using RG-213 coaxial cable.

Lowering the antenna to a height of 65' feet, as seen in Figure 23 and reducing the radiator lengths to obtain the best SWR, a value of 2.5/1 is achieved across the entire bandwidth, which is quite normal.

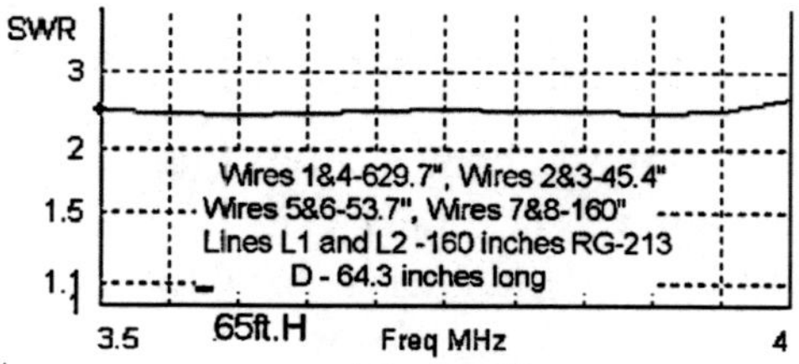

Figure 23

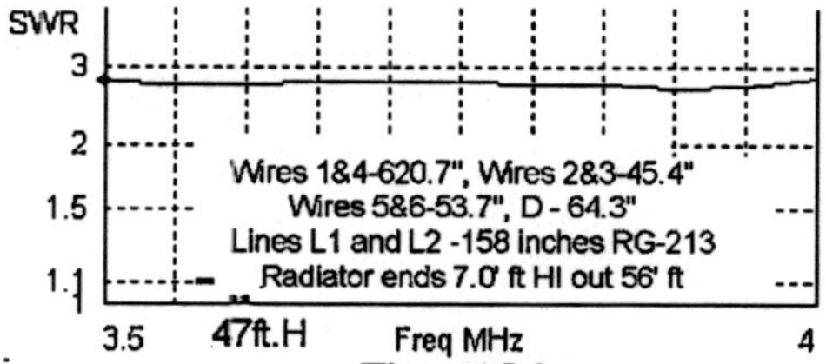

Figure 24

At a height of 47' feet, the radiator ends are 7.0' feet above ground and extend outward 56' feet, as seen in Figure 24 and the SWR is below 3/1, which is slightly better than a few of the other previous Diamond antennas.

The Far-field RF gain and loss chart of Figure 25 shows RF gains and losses pretty much in line with the previous antenna, making this another "Super-Diamond Antenna" with substantially zero loss, making it an extremely efficient antenna.

Far-field Gain & Losses at 89.7°			
MHz	3.50	3.75	4.0MHz
no loss	3.40	3.62	3.72dBi
w/ loss	3.39	3.59	3.59dBi
RF Loss	0.01	0.03	0.036dB
Coax-0.6dB/100'/10MHz Vf=0.66			

Figure 25

75-80m Normal Diamond Antenna with Loop Stubs

In the previous two Diamond Antennas the 2 Short-line Open-ended Stubs were connected in series and this gave them much higher voltage breakdown capabilities, but it also required twice the length than a single stub. By placing these stubs on Wires 5 and 6 it is possible to have them connected in parallel, instead of a series connection. This is only possible if the individual stubs are placed on adjacent Wires 5 and 6, as shown in Figure 26.

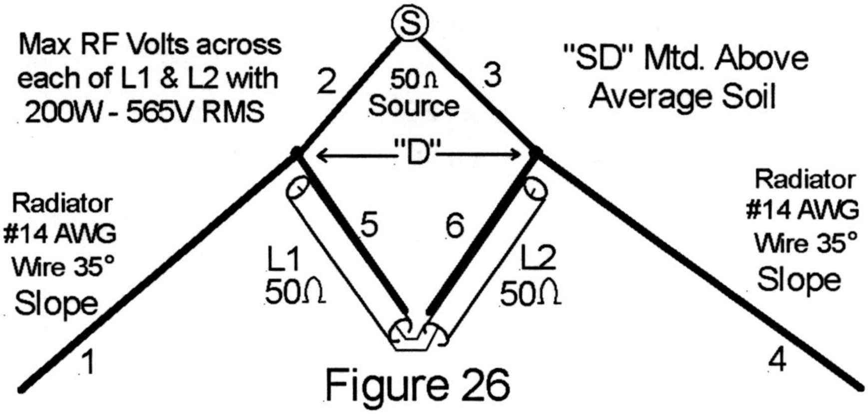

Figure 26

In order to do this, the center conductor of line L1 is connected to the outer braid at the lower end of line L2 and the center conductor of line L2 is connected to the outer braid of line L1. This effectively connects these two lines in parallel at the bottom ends of Wires 5 and 6. In the modeling process Wires 5 and 6 are specified as having a diameter of 0.32" inches in order to account for the coaxial cable size, but in reality #14 AWG wire is still used in the coaxial cable ends and are attached to these wires. By having these two lines in parallel, each of these lines are only required to be 50" inches in length and since there are two of them, that is equivalent to having a single line length of 100" inches.

These lines and wire lengths fit very nicely into a "Diamond" configuration, with Wires 2 and 3 being 46.85" inches long, whereas Wires 5 and 6 are each 50" inches in length, thus forming a very well-shaped Diamond loop.

This arrangement may cause some difficulties, if it is necessary to change either the wire or Stub lengths. If there is a need to increase the inductance of the loop, the upper ends of Wires 5 and 6 can be increased in length and the capacitive stub lengths can remain in its original length of 50" inches. On the other hand, if the resonant frequency of the loop needs to be increased, the upper ends of each of the two stubs can be shortened, still leaving the lengths of Wires 5 and 6 the same, without change.

With this arrangement and the antenna mounted at a height of 85' feet, the SWR plot of Figure 27 shows substantially the same SWR that was achieved with the previous Diamond configurations. This configuration is relatively small and easy to fabricate and uses a "D" length of 64.35" inches. Instead of having 160" inch long coaxial cables as part of the radiator which might cause some sagging, this antenna overcomes the possibility of this happening.

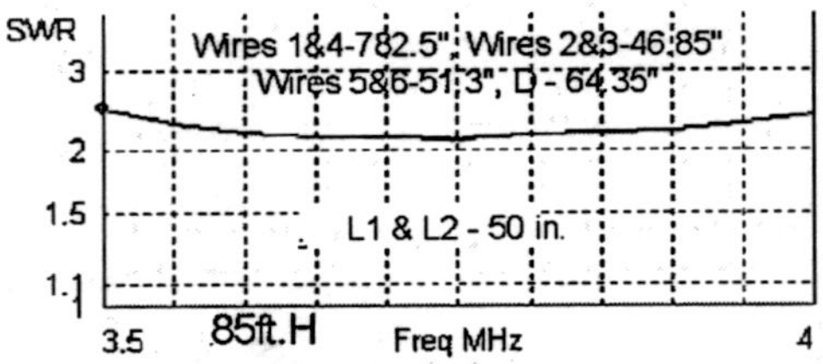

Figure 27

Once again, the normal 50/50 Balun is recommended between the Source and the transmission line and then the bottom ends of lines L1 and L2 should be secured to the downward transmission line with sufficient insulation between these items.

Far-field Gain & Losses at 55.9°			
85ft.H	3.50	3.75	4.0MHz
no loss	5.29	5.28	5.28dBi
w/ loss	5.29	5.28	5.28dBi
RF Loss	0.00	0.08	0.006dB
Coax-0.6dB/100'/10MHz Vf=0.66			

Figure 28

The Far-field RF gains and losses as shown in Figure 28 are typical of the previous antenna and the maximum radiation angle of 55.9° has not changed. The maximum loss of 0.08dB occurs at the center frequency and is very low at 0.08dB.

When this antenna is mounted at a height of 65' feet, the SWR still averages approximately 2.5/1, as shown in Figure 29, which has changed very little from other Diamond antennas in this chapter.

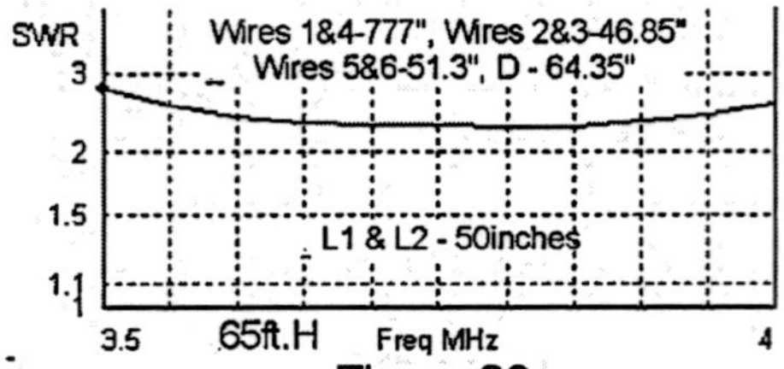

Figure 29

Lowering the antenna to 47' feet above ground will place the radiator ends at a height of 7.7' feet above ground and extend out 55.1' feet. The SWR obtained is slightly below 3/1 which appears to be the norm for these Diamond antennas as seen in Figure 30.

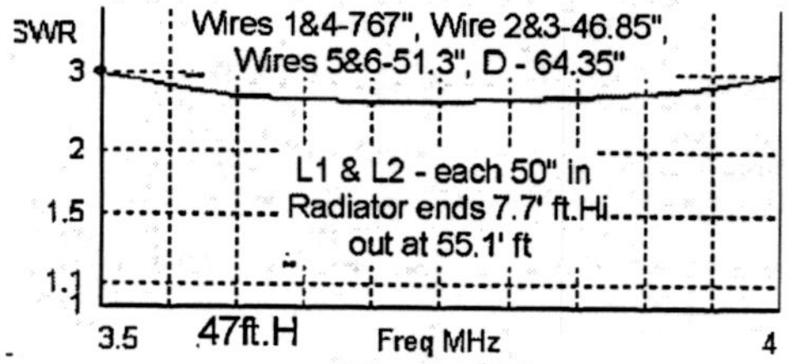

Figure 30

At a height of 47' feet the Far-field RF gains seem slightly greater and the losses unusually low at 0.01, 0.00 and 0.01dB across the 3.5 to 4MHz bandwidth, as shown in Figure 31. This is a Normal Diamond antenna having the source at the apex of the loop, which helps assure the best SWR. This is probably one of the best antennas, which substitutes the Stubs in order to eliminate a Fixed Capacitor and it is extremely easy to build and its performance is excellent and it surely can be considered a "Super-Diamond Antenna".

Far-field Gain & Losses at 89.7°			
47'ft.H	3.50	3.75	4.0MHz
no loss	3.42	3.67	3.79dBi
w/ loss	3.41	3.67	3.78dBi
RF Loss	0.01	0.00	0.016dB
Coax-0.6dB/100'/10MHz Vf=0.66			

Figure 31

75-80m Rotated Diamond Antenna with Loop Stubs

The Super-Diamond Antenna shown in the previous antenna had a normal configuration with the Source located at the Apex of the configuration. This antenna will have the loop rotated 180° so that the Source is located at the bottom and the Stubs are still on Wires 5 and 6, as shown in Figure 32.

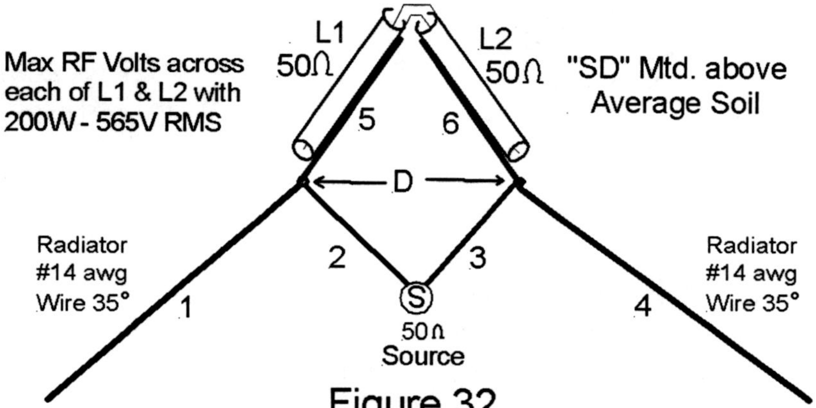

Figure 32

After rotating the antenna, it was found that the Diamond loop was under coupled from the radiator and therefore the common coupling Wires 2 and 3 had to be increased from 46.8" inches to 49.8" inches in order to increase the coupling. Additionally Wires 5 and 6 had to be increased from 51.3" to 52.8" inches in length. Once again, it was expected that these wires should have been shortened in order to compensate for the increase in length of Wires 2 and 3 but this is not the case and the wires actually had to be increased in length. The lengths of each of the two Stubs remain at 50" inches.

When this antenna is mounted at 85' feet the SWR is flat at an SWR of 2/1 across the band center and deteriorates slightly at the band edges, as shown in Figure 33. This Rotated version of the Diamond has consistently provided a slightly better SWR then the Normal version, it is believed that the additional Mutual Inductive Coupling that the normal version might have and some way inhibits having an SWR response as good as that of the Rotated version.

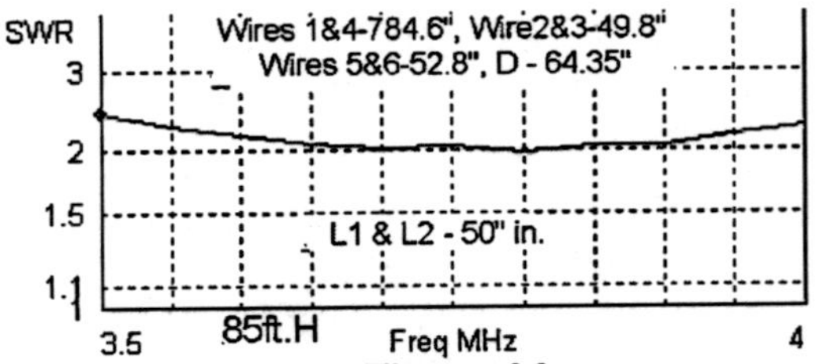

Figure 33

Far-field Gain & Losses at 56.8°			
MHz	3.50	3.75	4.0MHz
no loss	5.44	5.39	5.34dBi
w/ loss	5.44	5.39	5.33dBi
RF Loss	0.00	0.000	0.01dB
Coax-0.6dB/100'/10MHz Vf=0.66			

Figure 34

The Far-field RF gains as shown in Figure 34 are every bit as good as those shown for the Normal configuration and the coaxial cable losses are almost entirely nonexistent, with the worst reading being 0.01dB at 4MHz. This is a phenomenal antenna when it is rotated and it would be very difficult to find another Broad-Banded antenna that has characteristics even approaching those of this antenna.

At an antenna height of 65' feet, this antenna, as shown in Figure 35 has an SWR averaging above 2.5/1 across most of the bandwidth and this is about typical of the other Diamond configurations.

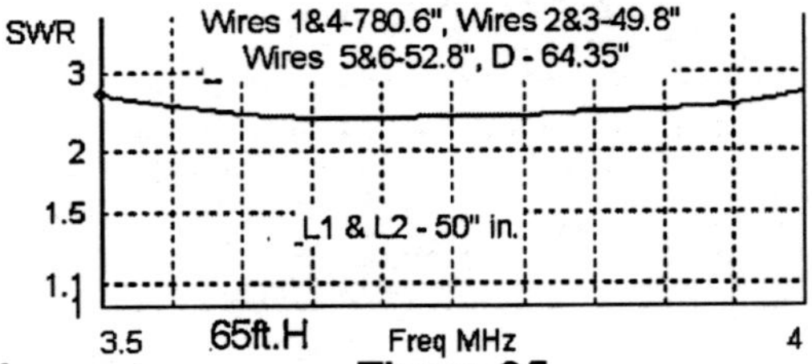

Figure 35

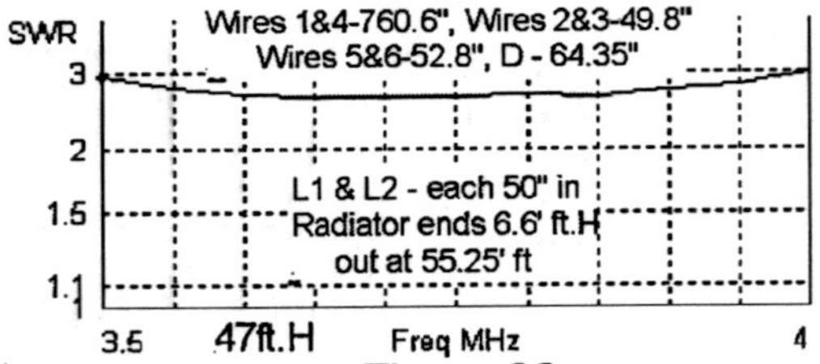

Figure 36

When mounted at a height of 47' feet, the radiator ends are only 6.6' feet above ground stretching out to 55.25' feet as shown in Figure 36. This was somewhat lower than desired and the antenna could be raised 6 to 10" inches for safety reasons. The SWR is somewhat below 3/1 across the entire bandwidth.

Far-field Gain & Losses at 89.7°			
MHz	3.50	3.75	4.0MHz
no loss	3.34	3.58	3.68.dBi
w/ loss	3.33	3.57	3.67dBi
RF Loss	0.01	0.01	0.016dB
Coax-0.6dB/100'/10MHz Vf=0.66			

Figure 37

The Far-field RF radiation gains, as shown in Figure 37 are quite good at 3.33, 3.57 and 3.67dBi across the 3.5 to 4MHz bandwidth and the coaxial losses across the entire bandwidth do not exceed 0.01dB.

Everything about this Diamond antenna seems to exceed most all other versions and it is difficult to find any faults with this antenna since it has all of the necessary qualities to be classified as a "Super-Diamond Antenna".

75-80m Rotated Diamond Antenna with Mid-Stubs

The Normal Hi Stubs Diamond antenna as shown in Figure 20 was modified to have the Diamond loop rotated 180° in which the Source is no longer at the Apex but is instead on the lower end of the Diamond, as shown in Figure 38.

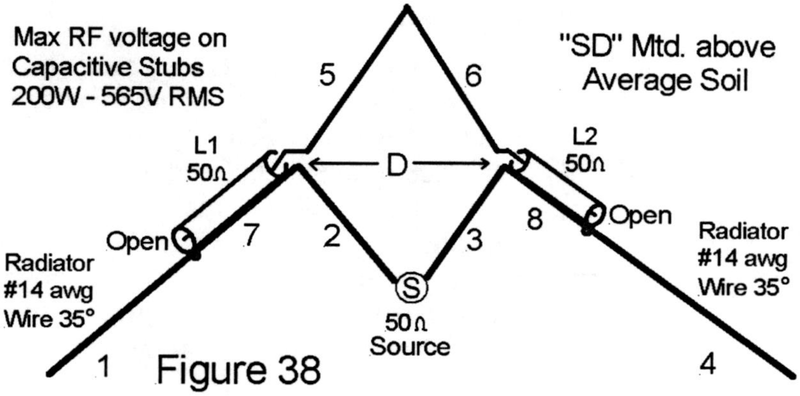

Figure 38

Keep in mind that this antenna is already in a rotated position and the relocated 160" inch long RG-213 coaxial open ended Stubs (L1 & L2) have been selected to this arbitrary length. Due to the Stub location it was necessary to decrease the lengths of Wires 2 and 3 from 49.8" to 45.4" inches, in order to increase the common coupling. The Mid Stubs (L1 & L2) lengths were selected arbitrarily and therefore Wires 5 and 6 had to be increased from 52.8" to 59.5" inches, in order to Re-resonate the Diamond and to bring the loop back to the proper resonant frequency.

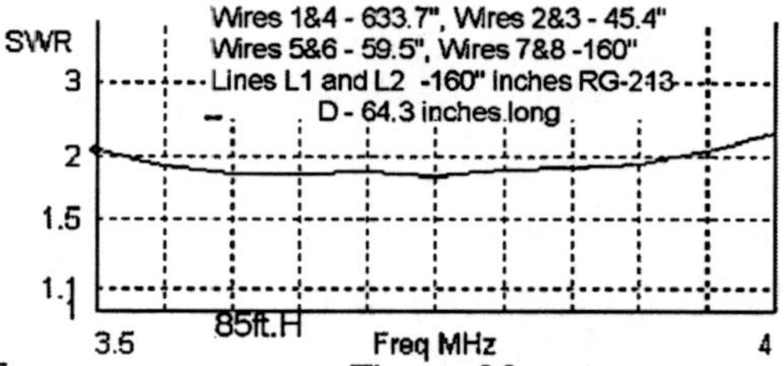

Figure 39

After choosing the final wire lengths, the SWR achieved when this antenna was mounted at a height of 85' feet are shown in Figure 39. This SWR is considerably better than many of the previous Diamond antennas and it has a value of less than 2/1 and it averages below this across the center of the bandwidth.

Since the Short-line Open-ended Stubs are mounted on the radiator Wires 7 and 8, the wire size was specified in the modeling of this antenna as 0.32" inches in diameter but these wires will still be #14 AWG wire and the coaxial stubs, which has substantially the same length, will be attached to these wires. Specifying the larger diameter Stub wire was necessary to ensure more accurate modeling of this antenna.

The Far-field RF radiation gain and loss chart of Figure 40 shows RF gains that are quite typical for an antenna height of 85' feet and the coaxial Stub losses are very low at 0.02, 0.02 and 0.06dB.

Far-field Gain & Losses at 57.6°			
MHz	3.50	3.75	4.0MHz
no loss	5.27	5.23	5.01dBi
w/ loss	5.25	5.21	4.95dBi
RF Loss	0.02	0.02	0.066dB
Coax-0.6dB/100'/10MHz Vf=0.66			

Figure 40

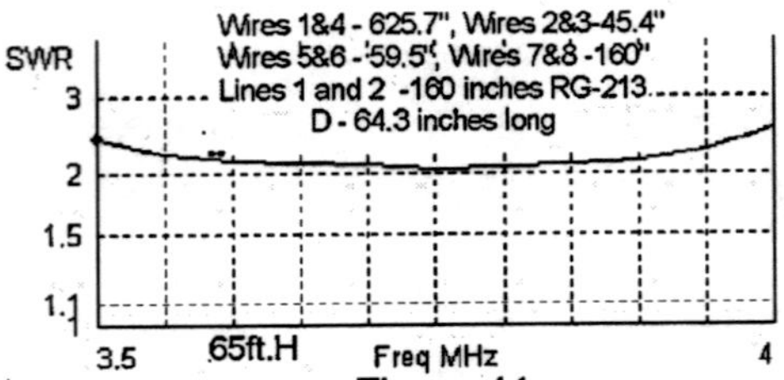

Figure 41

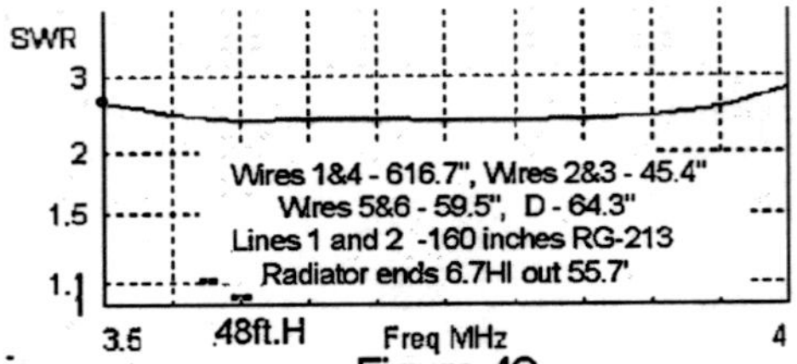

Figure 42

Reducing the antenna height to 65' feet produced a very good SWR averaging slightly over 2/1, as shown in Figure 41. These values are much better than other Diamond antennas that can only achieve these SWR values at a height of 85' feet; therefore this antenna is substantially better.

When this antenna is lowered to a height of 48' feet, as shown in Figure 42, the radiator ends will slope down to a height of 6.7' feet above ground, reaching out 55.7' feet. Although it is modeled at this height, it should be raised 5 to 10" inches higher. At this height of 48' feet, the SWR is considerably better than most Diamond antennas and it has a value of approximately 2.5. It is possible that the larger diameter of 0.32" inches, which is specified for Wires 7 and 8, is the reason for this improved SWR. It is well known that the use of larger diameter wire for the complete radiator can produce superior SWR figures.

Far-field Gain & Losses at 89.7°			
MHz	3.50	3.75	4.0MHz
no loss	3.39	3.55	3.53dBi
w/ loss	5.37	3.52	3.38dBi
RF Loss	0.02	0.03	0.05dB
Coax-0.6dB/100'/10MHz Vf=0.66			

Figure 43

The RF gains, as shown in the Far-field chart of Figure 43, shows that the gains are in-line with other Mid-Stub antennas, with values of 5.37, 3.52 and 3.38dBi, while the coaxial losses are extremely low.

Although this antenna might require slightly more work in placing the coaxial Stubs on the radiator, it has been shown to have excellent overall characteristics and as with all of the Diamond configurations which have all of the means of adjustment, thereby making this antenna in the classification of a "Super-Diamond Antenna".

75-80m Rotated Diamond Antenna with Hi-Stubs

The previous Rotated Mid-Stubs antenna was converted to a Hi-Stubs antenna by moving the two 160" inch 50 Ohm Short-line stubs from the bottom of Wires 5 and 6, up to the apex of the antenna, which is the top ends of Wires 5 and 6. These are Open-ended stubs, with their braids connected at both ends. In this arrangement, the stubs are connected in series and may require a length twice as long as that of a single coaxial stub.

Figure 44 shows this configuration and the stubs are hanging down from the Apex and since the stubs are 160" inches in length, they will hang lower than the source. It is recommended that a 50/50 Balun be used and when the transmission line is connected to the Balun, these two stubs can be attached mechanically to the 50 Ohm transmission line in order to prevent these lines from being blown during gusty winds and they can be readily secured.

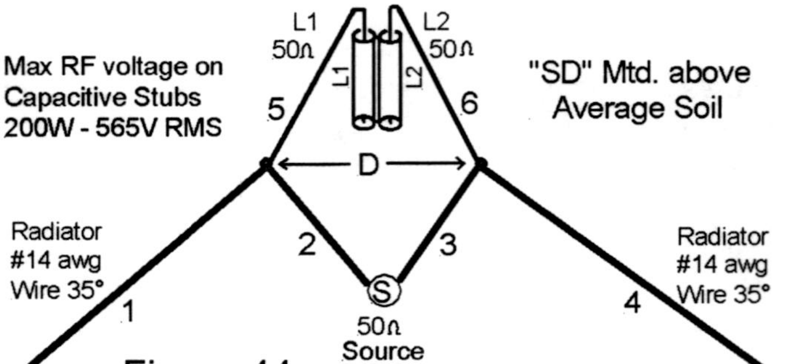

Figure 44

All of the wire diameters and lengths remained the same as the previous antenna, except that the radiator Wires 1 and 4 must be 788.6" inches in length. Each of the Open-ended stubs is still 160" inches in length.

At a height of 85' feet, this antenna has an SWR slightly lower than 2/1 across the bandwidth, as shown in Figure 45. This is substantially the same SWR as achieved in the previous Mid-Stubs Inverted V.

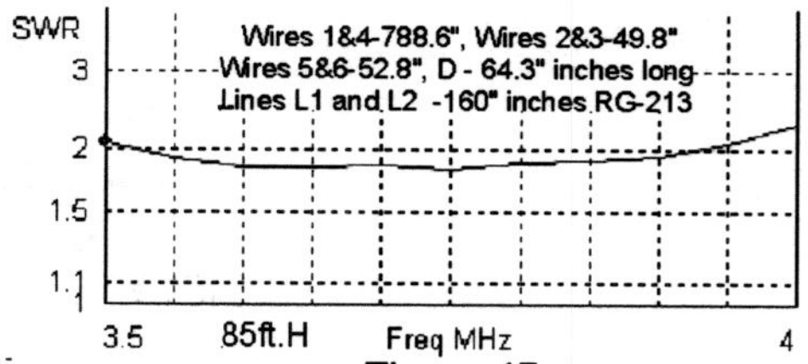

Figure 45

The Far-field RF gains, as shown in Figure 46, are all very typical of previous Diamond antennas with gains of 5.33, 5.21 and 5.14dBi. The RF losses are a bit higher, particularly at the 4 MHz end of the band, with losses of 0.11dB.

Far-field Gain & Losses at 55.8°			
MHz	3.5	3.75	4.0MHz
no loss	5.35	5.58	5.25dBi
w/ loss	5.33	5.21	5.14dBi
RF Loss	0.02	0.02	0.116dB
Coax-0.6dB/100'/10MHz Vf=0.66			

Figure 46

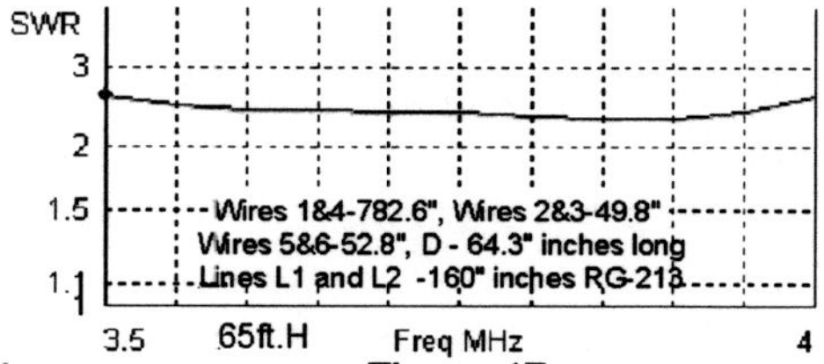

Figure 47

At a height of 65' feet, this antenna has an SWR averaging approximately 2.5/1 across the entire bandwidth and this is not as good as the previous antenna which had an SWR considerably better that was slightly over 2/1. It can only be concluded that the added diameter of the stubs when mounted on the radiator provides this improvement. When this antenna is lowered to a height of 47' feet, the radiator ends, when using a 35° slope, will be 6.6' feet high, reaching out 55.3' feet, as shown in Figure 48. It is best to raise this antenna 6" inches or slightly greater, out of reach. The SWR is well under 3/1 across the bandwidth, which is somewhat typical of other Diamond antennas.

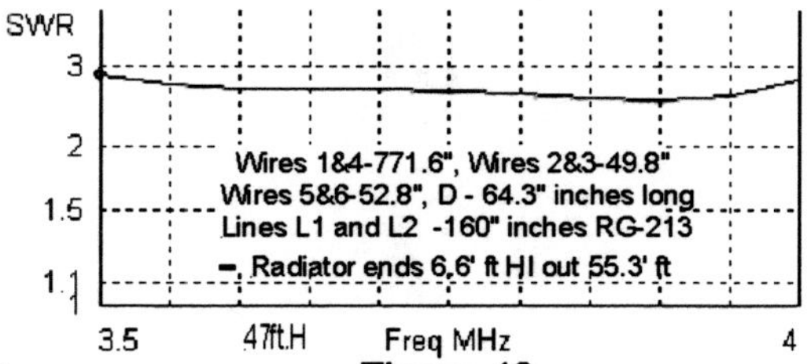

Figure 48

Although the previous Mid-Stubs antenna show a somewhat better SWR, it was more difficult to construct since the 160" inch long Stubs would have to be mounted on the radiator securely, whereas this antenna does not have quite as good of an SWR, is much simpler to build and the weight of the Stubs is directly in the center and at the Apex. On the previous antenna, these stubs will weigh heavily and cause some bending of the Inverted V radiator.

The Far-field RF gain and loss chart of Figure 49 shows that this antenna has good SWR values that are typical, but the coaxial losses are at 0.02, 0.03 and 0.14dB across the bandwidth. The value of 0.14dB is higher than we are becoming accustomed to in these Diamond antennas; it is a relatively low loss and hardly distracts from the antennas overall characteristics. It is one of the simplest Diamond stub antennas and it should be classified as a "Super-Diamond Antenna".

Far-field Gain & Losses at 89.7°			
MHz	3.5	3.75	4.0MHz
no loss	3.29	3.54	3.57dBi
w/ loss	3.27	3.51	3.43dBi
RF Loss	0.02	0.03	0.14dB
Coax-0.6dB/100'/10MHz Vf=0.66			

Figure 49

160m Rotated Diamond Antenna with Mid-Stubs

All of the Diamond antennas shown in this chapter have been modeled for the 75-80m band but there is a great need for Broad-Banded antennas in the 160m band. If we take a 75-80m Diamond antenna and scale it to the 160m band, the antenna will then be mounted at almost twice the height, such as an 85' foot antenna in the 75-80m band would have a height of 170' feet when operating at 160m. At these two different heights it would be above ground by the same percentage of a wavelength.

It is difficult for most Amateur radio operators to have tower heights that are over 100' feet and therefore this antenna will be mounted at a height of 90' feet. It cannot be lowered further since the radiator ends will already be approaching ground at a height of 9.7' feet, when using a radiator slope of 35°.

This antenna is a Mid-Stub antenna and the Stubs are intended to be mounted as part of the radiator, as shown in Figure 50. Unfortunately, the Stub lengths have increased dramatically when operating at 160m and this length is now 312" inches long for each of the L1 and L2 lines. If we were to use RG-213 coaxial cable for the stubs and we were to have lengths of 312" inches for each Stub to be mounted, this would put a heavy load on the supporting radiator wire which is specified as #14 AWG.

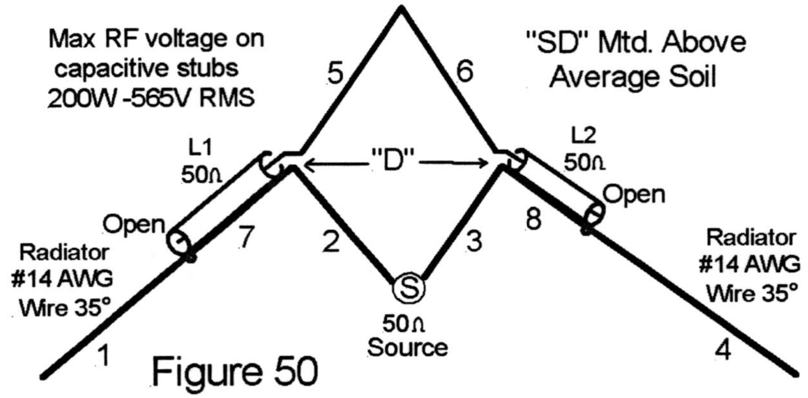

Figure 50

Since this diamond antenna will be scaled from the 75-80m band, to the 160m band, all of the wires and the coaxial Stub lengths will be approximately twice the length. The SWR plot of Figure 51 shows that the coaxial line lengths of L1 and L2 are each 312" inches in length, along with Wires 7 and 8, on which these lines are mounted. Wires 2 and 3 are now 80.3" inches long, while Wires 5 and 6 are 104.3" inches in length, as shown in Figure 51. The radiator lengths have then been trimmed to 1221.4" inches in length and they extend down to a height of 9.7' feet above ground reaching out 109' feet.

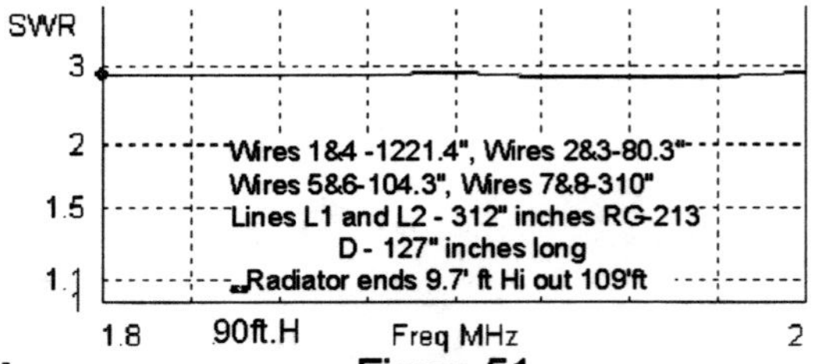

Figure 51

The SWR plot of Figure 51 indicates an SWR of slightly less than 3/1 across the entire band mounted at a height of 90' feet and this is the best that can typically be obtained. Nevertheless, Transceivers which have an internal Antenna-Tuner can function properly at this SWR and improve the match to the transmitter and the receiver.

The Far-field gain and loss chart of Figure 52 shows gains of 3.92, 3.92 and 3.74dbi across the 160m band and this is slightly better than you might obtain if this antenna was rescaled back to the 75-80m band, at which it would be mounted at a height of 45'-47' feet. At this height previous Diamond antennas would have a typical 3/1 SWR value. Therefore, it seems that this SWR is the best obtainable with Diamond antennas in the 160m band, however operation of an Antenna-Tuner, whether internal or external, will greatly improve the SWR.

Far-field Gain & Losses at 89.9°			
MHz	1.8	1.9	2.0MHz
no loss	3.96	3.96	3.91dBi
w/ loss	3.92	3.92	3.74dBi
RF Loss	0.04	0.04	0.176dB
Coax-0.6dB/100'/10MHz Vf=0.66			

Figure 52

The coaxial losses, as shown in Figure 52, are 0.04, 0.04 and 0.176dB across the entire 160m bandwidth, from 1.8 to 2MHz. These are all acceptable losses and this antenna has all of the qualities necessary to be classified as a "Super-Diamond Antenna".

160m Normal Diamond Antenna with Fixed-Capacitor

The previous Diamond antenna as shown in Figure 50 was a Rotated Mid-Stub configuration and it was converted to a Basic Normal Diamond version that utilized a Fixed-Capacitor to resonate the Diamond loop. In doing this, the Source is now located at the apex of the radiator and the resonating Capacitor, which had to be altered to 450pF due to scaling from the 75-80m band, is now located at the bottom end of the Diamond at Wires 5 and 6, as shown in Figure 53.

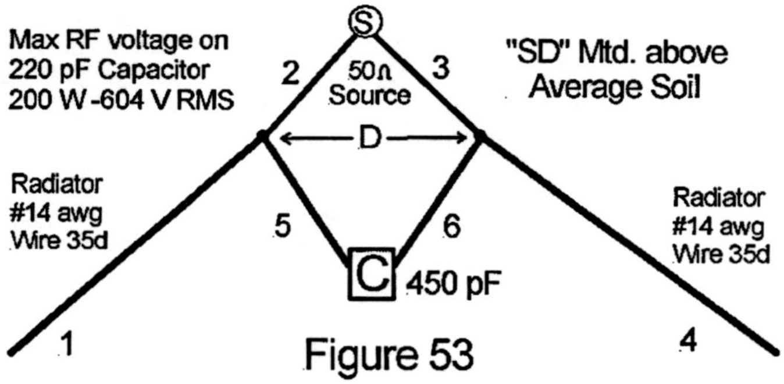

Figure 53

In this arrangement Wires 7 and 8 have been eliminated and the main radiator ends extend down to a height of 11.4' feet above ground at 110' feet. Figure 54 shows an SWR slightly higher than 3/1, which can probably be improved if an Internal Antenna-Tuner, is used.

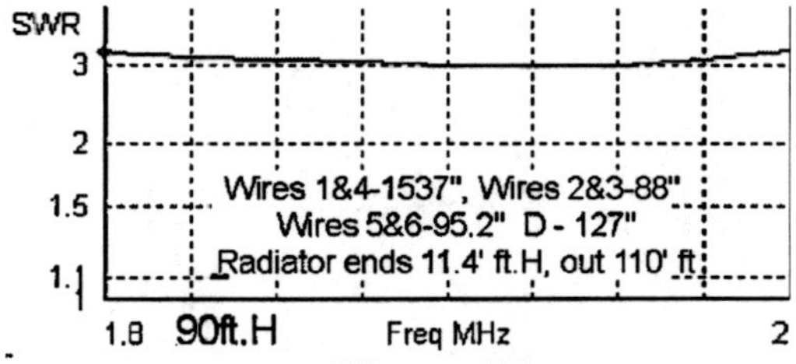

Figure 54

The Far-field gain chart of Figure 55 shows gains of 3.82, 3.99 and 4.0dBi across the 160m bandwidth and at 1.9MHz, the angle of maximum radiation is 89.9°.

Far-field Gain & Losses at 89.9°			
MHz	1.8	1.9	2.0MHz
RF gain	3.82	3.99	4.00dBi

Figure 55

If you are wondering what voltages there might be on the 450pF Capacitor, the chart of Figure 56 shows RF voltages of 149, 605 and 556V rms. These voltages are listed for a transmitter power of 200 Watts and with a power level of 800 Watts these voltages will be twice as high.

This is probably the simplest of the Diamond antennas, since no Stubs are used, however you must contend with the tolerance of the Capacitor and be aware that some trimming of the lower ends of Wires 5 and 6 probably have to take place in order to obtain asymmetrical SWR. As long as you have 5% Capacitor tolerance, this trimming may not be needed.

RF volts on capacitorwith 200 W			
MHz	1.8	1.9	2.0MHz
RF volts	149	605	556rms

Figure 56

This antenna utilizes the ease of adjustment available in the Diamond configuration, is very simple to build and it should perform well. It is to be considered a "Super-Diamond Antenna".

160m Rotated Diamond Antenna with Loop-Stubs

This antenna is a rotated Diamond configuration which has the Source at the lowest end of the Diamond and having 2 Short-line Open-ended coaxial Stubs. Both of these stubs are very short and although each is mounted on a separate wire, these stubs are actually in parallel, permitting the length of each stub to be 98" inches in length, as shown in Figure 57.

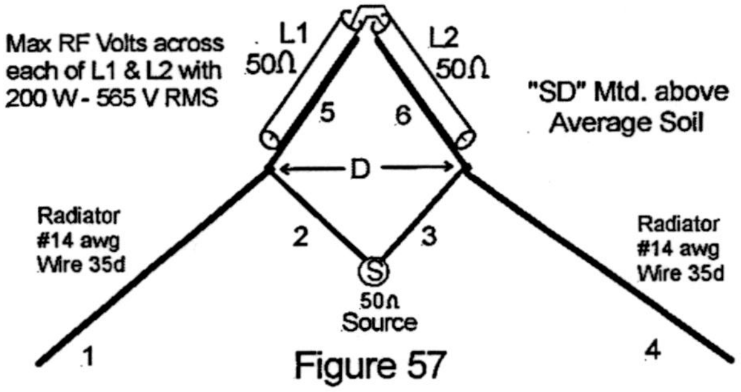

Figure 57

On close examination of the Diamond loop, you will find that the very tops of the Stubs are connected in an unusual way in order to have them connected in parallel, while mounted on different wires. Stub L1 has its center wire conductor connected to the braid of Stub L2 and the center conductor of L2 is connected to the braid of L1. When these two Stubs are connected in parallel the resulting capacitive reactants of these stubs at 1.9 MHz will be equivalent to approximately 450pF, which is the desired value.

These two stubs are primarily supported by the mounting Apex and they are quite short in length using RG-213 coaxial cable, which has a solid dielectric and much higher breakdown voltage than cables with Foam insulation.

The SWR obtained, as shown in Figure 58 is relatively close to a value of 3/1 across the entire 160m bandwidth, when the antenna is mounted at a height of 90' feet. This is consistent with the other 160m Diamond antennas presented so far.

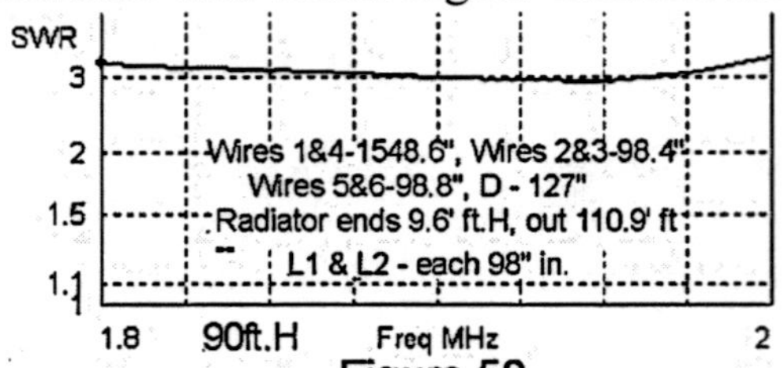

Figure 58

The Far-field gain and loss chart of Figure 59 shows gains of 3.71, 3.84 and 3.81dBi across the entire 1.8 to 2MHz band and at a radiation angle of 89.9°. This is very good gain for this antenna height for the 160m band and the coaxial losses are 0.00, 0.00 and 0.02dB. This is an unusually low loss, but considering the short lengths of the Open-ended coaxial stubs, this lower loss should be expected.

Far-field Gain & Losses at 89.9°			
MHz	1.8	1.95	2.0MHz
no loss	3.71	3.84	3.83dBi
w/ loss	3.71	3.84	3.81dBi
RF Loss	0.00	0.00	0.02dB
Coax-0.6dB/100'/10MHz Vf=0.66			

Figure 59

Remember, these two coaxial stubs are in parallel compared with other Diamond antennas. In order to have a balance, 2 Stubs are used and they are in series, thus giving an RF voltage rating that is twice as high as that of a single Stub or 2 Stubs in parallel.

Of all of the antennas shown, this is the simplest and least expensive and it has all the qualities that make it an excellent selection and it is surely a "Super-Diamond Antenna".

160m Normal Diamond Antenna with Loop-Stubs

The previous antenna was a Rotated-Diamond, with Loop-Stubs and it was changed to a Normal Configuration, where the Source is at the apex of the Diamond and the parallel Loop-Stubs are still attached to Wires 5 and 6, as shown in Figure 60.

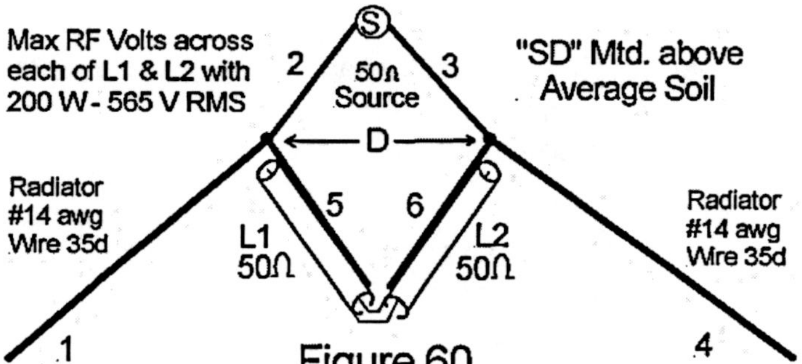

Figure 60

In order to rotate the common Diamond Loop, the amount of common coupling lengths of Wires 2 and 3 require changes. If you rotate The Diamond loop from a Normal configuration to a Rotated version, the coupling between the Diamond Loop and the radiator will be under coupled and Wires 2 and 3 must be increased in length to achieve greater coupling. In doing this, other wires lengths may also require readjusting.

On the other hand, rotating the Diamond Loop from a Rotated configuration to a Normal configuration, the coupling will increase and Wires 2 and 3 must be reduced in length. This particular antenna is a Normal configuration and the common coupling Wires 2 and 3 have been changed from 98.4" inches, to a shorter length of 86.1" inches.

In order to have the Diamond Loop re-resonate to the most desirable frequency in the 160m band pass, Wires 5 and 6 had to be increased from 98.8" inches to a length of 101.5" inches. Due to the shortening of Wires 2 and 3, the ends of each radiator had to be reduced from 1548.6" inches down to 1543" inches. With these changes the restraining cord "D" had to be changed from 127" inches in length down to a value of 116" inches.

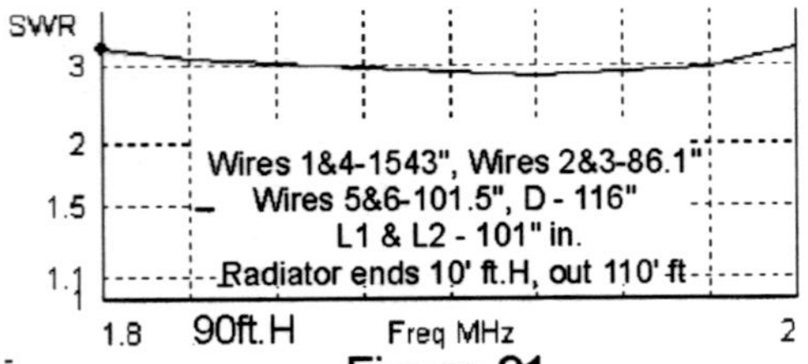

Figure 61

The SWR plot of Figure 61 shows an SWR of 3/1 after these changes have been made. This is a typical SWR for all of these 160m antennas that are mounted at 90' feet above Average-soil.

The Far-field RF gain and loss chart of Figure 62 shows gains of 3.95, 4.11 and 4.17dBi across the entire 160m bandwidth. This is an improvement over the Rotated configuration of Figure 57, which shows gains of 3.71, 3.84 and 3.81dBi.in the chart of Figure 59. This is due entirely to the fact that Wires 2 and 3, which form a part of the radiator and they have the highest RF currents, thus producing the greatest gain at the highest point on the Diamond Loop. In contrast, the Rotated version has Wires 2 and 3 down at a much lower height and since they are still part of the radiator, the RF antenna gains are lower.

Far-field Gain & Losses at 89.9°			
MHz	1.8	1.9	2.0MHz
no loss	3.95	4.12	4.19dBi
w/ loss	3.95	4.11	4.17dBi
RF Loss	0.00	0.01	0.02dB
Coax-0.6dB/100'/10MHz Vf=0.66			

Figure 62

The coaxial losses were measured during the modeling process and they have been noticeably the same as the previous Rotated version, at values of 0.00, 0.01 and 0.02dB across the entire 160m bandwidth. This is about as low as you might expect and coupled with the improved RF gains, this antenna overwhelmingly meets the requirements of a "Super-Diamond Antenna".

In order to have a reminder of the need for Broad-Banding of the antennas in the 160m band, a "Basic Inverted V" antenna was modeled at a height of 90' feet, using a radiator slope of 35° with #14 AWG wire. Its resonant point, which is the best SWR, was centered at 1.9MHz as shown in Figure 63.

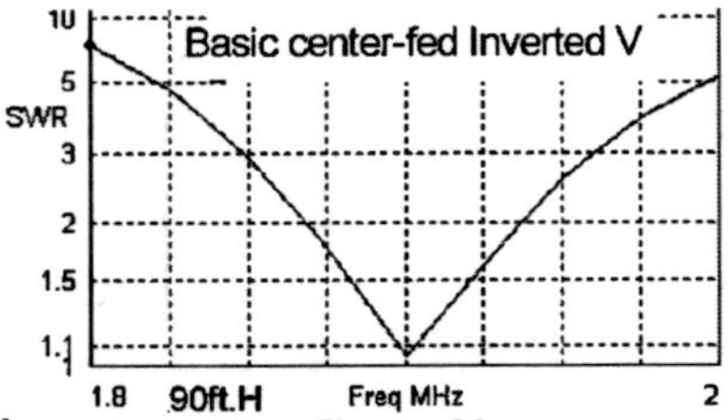

Figure 63

If you were only interested in frequencies towards the center of the bandwidth, this basic Inverted V might seem adequate, however even in that area; there will be a need for constantly retuning the Antenna-Tuner. The SWR varies at the band ends from an SWR of 5 up to approximately 8 or 9/1 and in the SWR above 3/1 it will require an external Antenna-Tuner, which will require constant retuning each time the frequency of operation is changed.

At a height of 90' feet, at least 100' feet of coaxial cable will be required as a transmission line and if you had a basic SWR of 9/1 at the lower 160m band edge; you can expect extremely high losses in this transmission line and as stated previously, the Antenna-Tuner, when located in the radio room, you cannot in any way remove these losses, which might amount to over 1/4 of your RF power to be dissipated in this line.

Numerous versions of "Super-Diamond Antennas" have been presented in this chapter, so there are a large variety of the antennas to choose from. Each of them has their own unique qualities and characteristics; you can choose the one that best suits your needs. The entire group has been developed by a modeling program, with the utmost care taken to provide the greatest accuracy. These antennas are at this stage completely new and State of the Art, but as you read the details of their specifications and various advantages, I believe that one of these will be your antenna of choice.

Super Max-Dipole Antennas
By Edward J. Shortridge W4JOQ

Chapter 5
Super-Dual Inverted-V Antennas

Introduction

Inverted V antennas have not been a very good choice for use in the 75-80m and 160m bands due to their highly restricted SWR bandpass, which has an SWR well over 3/1 over a good portion of the total bandwidth and greatly increasing towards the band edges. This condition requires constant Antenna-Tuner use as the frequency of operation is altered. The typical Antenna-Tuner of most transceivers have their operation limited to an SWR of approximately 3/1, while the antenna SWR may increase to extreme values at the band edges. Each time the frequency of operation is altered, the Antenna-Tuner must be used and this can become quite aggravating. On the other hand, if the Inverted V can be Broad-Banded to the point where better than a 2/1 SWR can be obtained across the entire bandwidth, this constant retuning may be avoided and the transmission line losses can be greatly decreased.

Throughout the previous book titled "Super-Dipole Antennas" and through the first four chapters of this book "Super-Max Dipole Antennas", all efforts have been concentrated on improving the SWR by means of the "Critical-Coupling Method" of achieving improved SWR values across the entire bandwidth. Unfortunately, good SWR values can be obtained only at the higher antenna heights and as the antenna is lowered, the SWR deteriorates significantly, some worse than others.

Still this Dual-Inverted V Broad-Banding method is vastly improved over the Non-Broad-Banded antennas but as an alternative to overcome the "Critical-Coupling Method" limited capabilities would be highly desirable and in this chapter an alternative Broad-Banding method will be presented.

Dual-Inverted V Broad-Banding

A method of Broad-Banding will be presented, which utilizes two Inverted V antennas mounted at right angles to each other, but fed by a single Source. These two antennas are isolated from each other since they are at right angles and they are only coupled by their Source.
One Inverted V is resonated near the lower end of the desired bandwidth, while the other Inverted V is resonated towards the higher frequency end. In doing this, their common Source has overlapping impedances that tend to compensate for each other, producing a very low SWR across the entire bandwidth. Each of these Inverted V radiators may have their lengths readjusted at any time so that they may provide the very best SWR. Each Inverted V has a Source impedance of approximately 50 Ohms and when the two Inverted V radiators are connected in parallel at the Source and then the most desired resulting impedance has increased to approximately 60 Ohms, thus requiring a short length of "Series-section Matching Line".

Since each of the Inverted V radiators do not rely on any Mutual Coupling, this low SWR is maintained at all antenna heights, which is in direct contrast to the "Critical-Coupling Method" results, in which the SWR suffers badly at low antenna heights. Surprisingly, the SWR values achieved with the Dual-inverted V's are far superior at all antenna heights.

However, these Dual-Inverted V antennas are not necessarily suited to everyone's use, since they take a considerable amount of space for the mounting area. Nevertheless, they are very adaptable to all antenna heights, down to having a horizontal radiator mounted at 8' feet above ground and they can still retain very low values of SWR across the entire bandwidths in both the 75-80 and 160m bands.

75-80m Dual-Inverted V at a 45° Radiator Slope

This first Dual-Inverted V antenna utilizes an antenna slope of 45°, which generally creates difficulty in obtaining a reasonable SWR throughout the bandwidth, however in order to help the SWR, a wire diameter of 0.32" inches OD, is used as shown in Figure 1. **See "Velocity Factor" at end of page.**

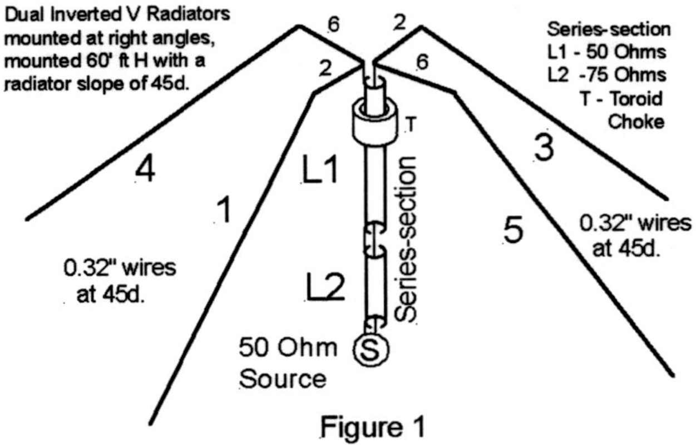

Figure 1

It was desired that this antenna would match directly to a 50 Ohm Source but instead, the best overall SWR required a source impedance of 60 Ohms. Therefore a "Series-section Matching Line", which consists of lines L1 and L2, which have line impedances of 50 and 75 Ohms respectively had to be used. These lines are relatively short, not critical at all in length since the match is only from 62 to 50 Ohms, resulting in very little loss. The lengths are 251.8" inches and 82" inches, which also acts as part of the downward transmission line.

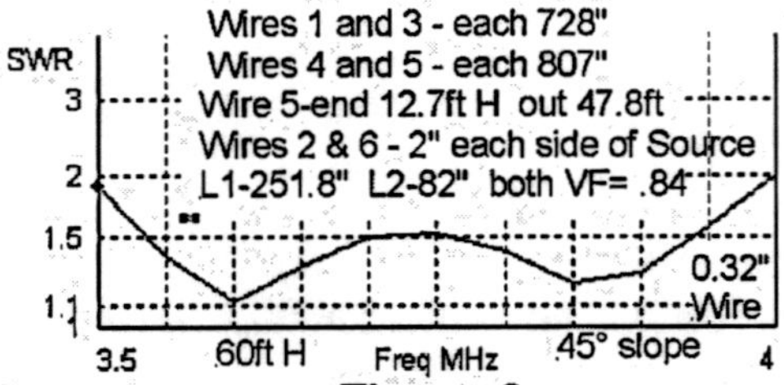

Figure 2

After adjusting the length of the radiator ends in accordance with those values shown in Figure 2 an excellent SWR plot resulted, which is generally below an SWR of 1.5/1, except for the extreme edges of the bandwidth. At a slope of 45° and at a height of 60' feet, this is an extremely good SWR, which would be difficult to duplicate by any other Broad-Banding method.

The Far-field RF radiation gains of Figure 3 indicates values of 4.12, 4.50 and 4.71dBi, which is very good for an antenna of this type mounted at a height of 60' feet. The coaxial line losses are 0.17, 0.10 and 0.08dB across the 75-80m bandwidth. These are relatively low losses and a good portion of these must be assigned to the downward transmission line. Overall, this is a unique antenna, which will perform exceedingly well.

Far-field Gain & Losses at 89.8°			
MHz	3.5	3.75	4.0MHz
W/o loss	4.29	4.60	4.79dBi
With loss	4.12	4.50	4.71dBi
Coaxloss	0.17	0.10	0.08dB

60ft. H Figure 3 45° slope

When more typical wire diameters, such as #14 AWG are used, as shown in Figure 4, the SWR has deteriorated in the center of the bandwidth from an SWR of 1.5 to a value of 1.8. This is still a much better SWR that can be obtained by any of the previous shown Broad-Banded configurations at a 60' foot height.

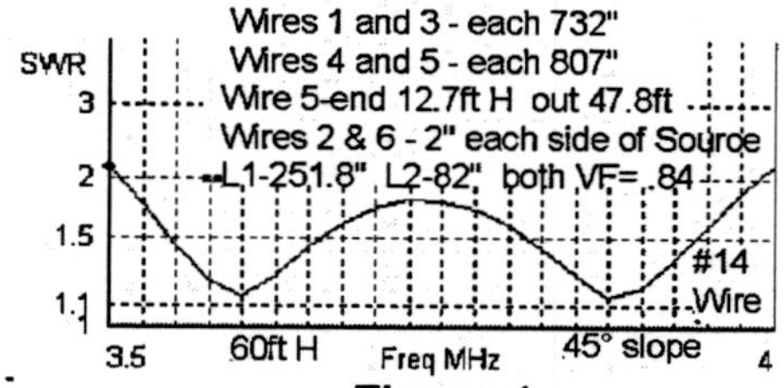

Figure 4

In order to determine the effectiveness of the "Series-section Matching Line", it was removed and a 50 Ohm Source was placed directly at the top of the Dual Inverted V configuration at Wire 2. The resulting plot of Figure 5 shows only a modest deterioration of the SWR from a value of 1.8/1 at the band center, to a value of approximately 2/1. On other similar antennas mounted at different heights, these will have somewhat different effects.

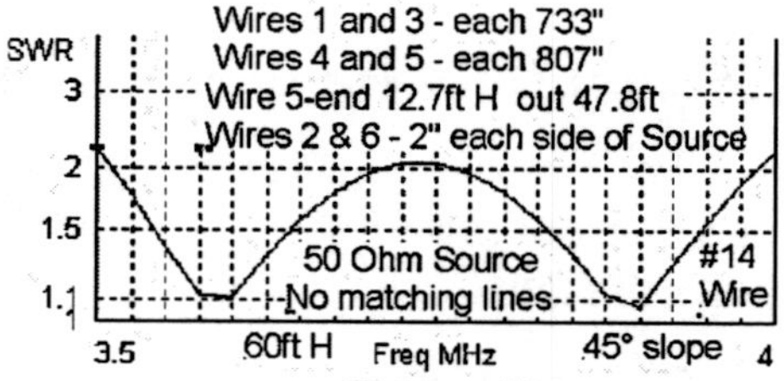

Figure 5

When either of these antennas utilize a large diameter wire such as the 0.32" inch OD, or a #14 AWG wire, there is little question that the SWR values obtained are far superior to those obtained when utilizing the "Critical-coupled Broad-Banding Method", such as that used in the "Diamond" configurations.

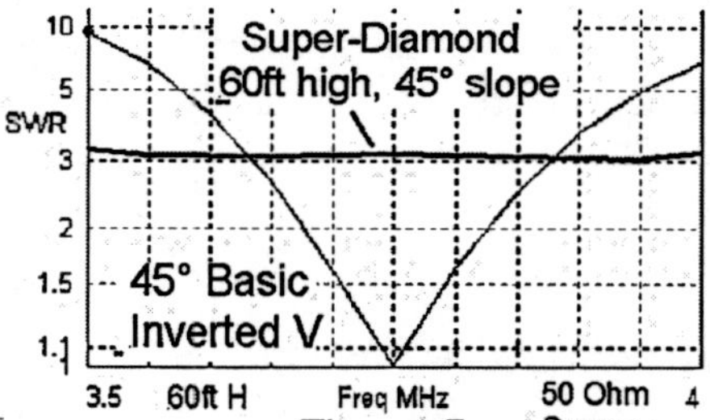

Figure 5a

In order to show the contrast, a Normal Diamond Inverted V was modeled having a mounting height of 60' feet, with a radiator slope of 45° and the resulting SWR plot of Figure 5a shows the resulting SWR, which has a flat 3/1 slope SWR. Comparing this Figure 5a plot to that of Figure 5, there is little question that the "Dual-Inverted V" is far superior and its use should be promoted as well as utilized throughout the Ham radio community.

A portion of the SWR of Figure 5a shows the same "Inverted V" in which the "Diamond Broad-Banding" has been removed and a basic 45° single "Inverted V" is used. Now a real comparison can be made to show that the basic plot of Figure 5a really needs Broad-Banding capability and then how improved the plot can be when the "Dual Inverted V" configuration is utilized, along with the "Series-section Matching", as shown in Figure 4. There is little question that there are huge improvements.

If you believe that "Series-section" Broad-Banding is not necessary and the plot of Figure 5a is adequate for your needs, then the Far-field RF gains of Figure 6 may be of interest, with gains of 4.17, 4.49 and 4.63dBi across the entire 3.5 to 4MHz bandwidth.

Far-field Gain at 89.8°			
MHz	3.5	3.75	4.0MHz
RF gain	4.17	4.49	4.63dBi

60ft. H 45° slope

Figure 6

Velocity Factor - If you can find the exact coax impedance you require, but the Velocity Factor is unavailable you can select a new and different one; however you must modify the coaxial cable length. In order to make this change, divide the existing coax length by the existing Velocity Factor and multiply it by the newer one, regardless of the operating frequency. Try to keep cable diameters, insulations, and center-wires the same as previously. There are plenty of 50 Ohm coax to choose from, but when you require a 75 Ohm value, there are much fewer, so it becomes necessary to consider a different Velocity Factor.

When the "Series-section Matching Lines" are used, you must realize that these coaxial cables are an unbalanced line and in order to prevent RF currents from flowing down on the outside braid, it may be necessary to provide isolation. Since the lengths of the lines are important, their lengths cannot be extended or reduced and the simplest way is to provide a very short portion of the upper part of line L1 with a smaller diameter coaxial cable that can be wound on the Toroid Core, with just sufficient amount of inductive reactants to prevent these currents. This small length of coaxial cable on the winding must be part of line L1 having a total specified length of 249.8" inches in length.

If a regular 50/50 Balun is to be used for isolation, it would either have an unknown length of coaxial cable or #14 AWG wires used and this would extend the length of line L1 and this would seem like a totally unacceptable arrangement.

75-80m Dual-Inverted V at a 35 ° Radiator Slope

By reducing the antenna slope from 45° to 35°, further improvements can be made in the SWR bandwidth and also the antenna can be lowered while still keeping the radiator ends well out of reach.

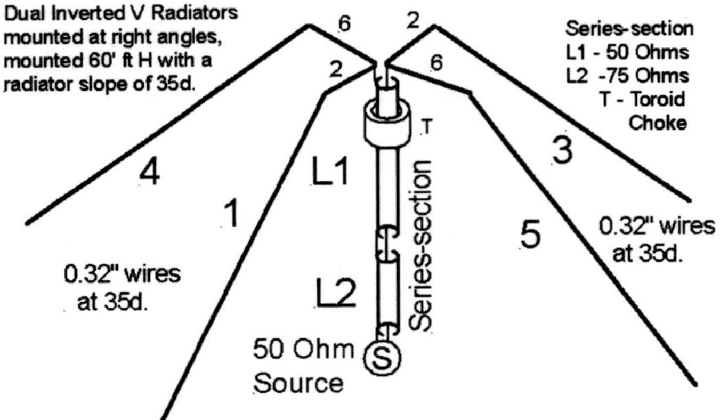

Figure 7

The same configuration of the previous antenna is used and only the antenna slope of 35° impacts this change, with the heights remaining at 60' feet, as shown in Figure 7.

The SWR plot of Figure 8 shows some improvement, both in the center of the bandwidth and at the band edges, but although these improvements are small it becomes more difficult to obtain improvements, when the SWR is below 1.5/1, as is the case with this plot.

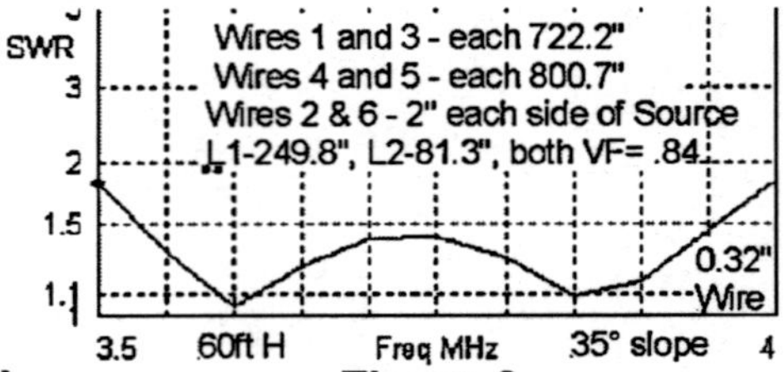

Figure 8

Far-field Gain & Losses at 89.8°			
60ft.H	3.5	3.75	4.0MHz
W/o loss	5.08	5.27	5.37dBi
With loss	4.92	5.17	5.30dBi
Coaxloss	0.16	0.10	0.07dB

Figure 9 35° slope

Due to the reduced radiator slope of this antenna, the Far-field RF gains have increased to values of 4.92, 5.17 and 5.30dBi across the entire 3.5 to 4MHz bandwidth, as shown in Figure 9, compared to the 4.12, 4.50 and 4.71dBi of the previous antenna, both of which are mounted at a height of 60' feet.

The antenna, as shown above in Figures 7 and 8 utilizes a large diameter wire of 0.32" inches OD and this has been changed to a #14 AWG wire size, as shown in the SWR plot of Figure 10. This has resulted in an improved SWR across the entire center of the bandpass, deteriorating to an SWR of 2/1 at the band edges. Whereas, when the radiator slope of 45° was used in the previous antenna, the center bandpass SWR was approximately 1.8/1.

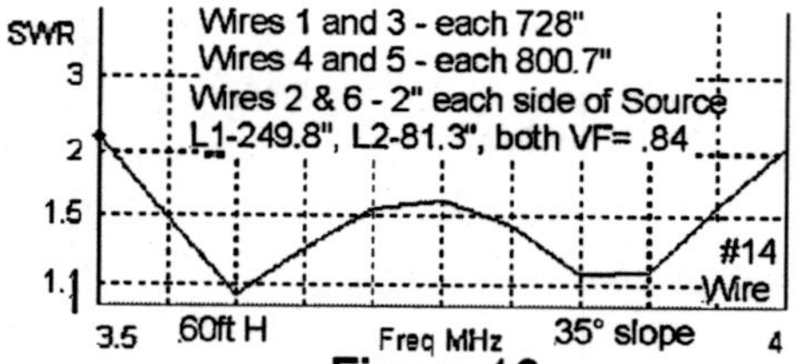

Figure 10

The height of 45' feet is the lowest that this antenna can be mounted since the 35° radiator slope must provide the radiator ends a minimum clearance of 7' feet above ground. In Figure 11, the ends of Wires 1 and 3 are at a height of 10.43' feet extending out 49.6' feet, while Wires 5 and 6 are at a height of 7.5' feet, extending out 54.4' feet.

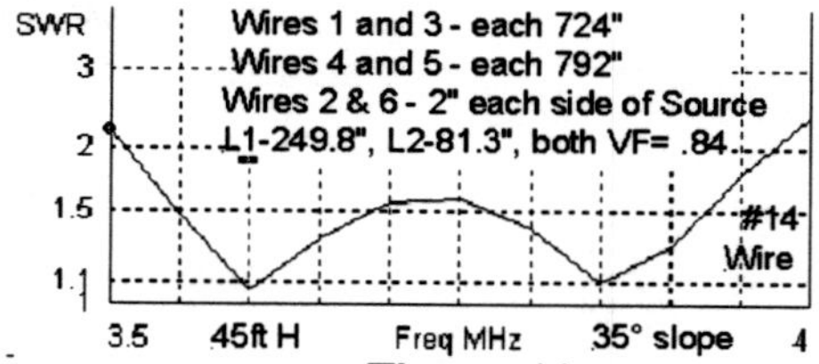

Figure 11

This antenna when mounted with a radiator slope of 35° has excellent SWR bandwidths, low coaxial losses in the "Series-section Matching Lines" and has excellent Far-field RF gains and it should be classified as a "Super-Dual Inverted V Antenna". Compare it to the best Broad-Banded antenna that might be available, you will find that this antenna gets the highest marks.

75-80m Dual-Inverted V mounted at 25'ft and 8'ft High

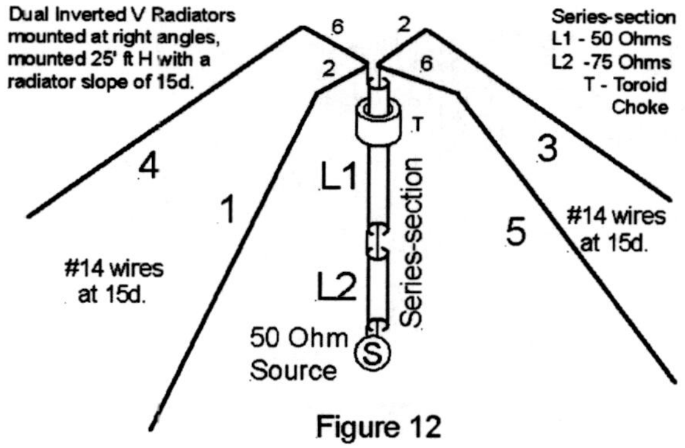

Figure 12

In order to demonstrate the full capabilities of the "Dual-Inverted V", the previous antenna was lowered to a mounting height of 25' feet above Average-soil and a #14 AWG radiator wire is used. In order to keep the radiator ends well out of reach, the antenna slope had to be reduced to 15°, as shown in Figure 12.

The lower ends of Wires 1 and 3 are now at a height of 9.5' feet and extending outward, while Wires 4 and 5 are at a height of 8.5' feet. After the wires have been adjusted in accordance with Figure 13, an excellent SWR plot is shown across the center of the bandwidth at a value of slightly above 1.5/1, while the band edges have an SWR of 2/1.

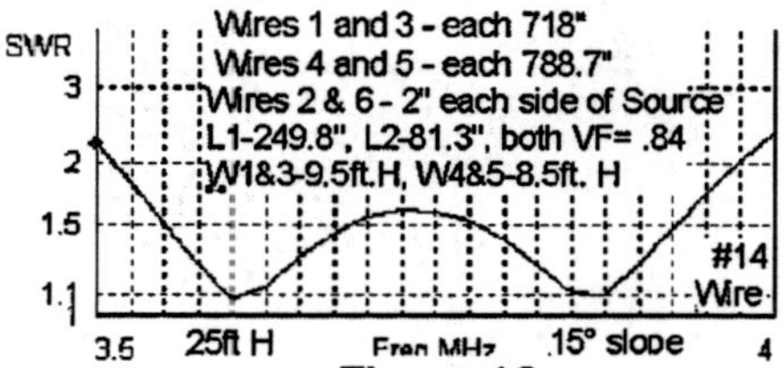

Figure 13

This is a terrific SWR for any antenna mounted this low and having a "Dual-Inverted V" is a prime reason for achieving this excellent SWR. And although the antenna has an Apex of only 25' feet, the Far-field RF gains of Figure 14 are still quite reasonable at 1.69, 2.30 and 2.77dBi. Of course, the maximum radiation is at practically 90° from horizontal and nothing can be done to improve this, except for mounting the antenna at heights of approximately 80' feet or higher.

Far-field Gain & Losses at 89.8°			
25ft.H	3.5	3.75	4.0MHz
W/o loss	1.87	2.39	2.83dBi
With loss	1.69	2.30	2.77dBi
Coaxloss	0.18	0.09	0.06dB

Figure 14 15° slope

In order to have a common height for all of the 4 radiator ends, they were adjusted to a height of 7.6' feet above ground. This is shown in Figure 15 and this required some very minor changes in the radiator wire lengths in order to maintain the best SWR across the entire bandwidth. In doing this, the radiators slopes have been altered slightly from the 15° of the previous configuration, but it can greatly simplify the mounting of this antenna and as you can see the SWR is substantially the same as the previous Figure 13.

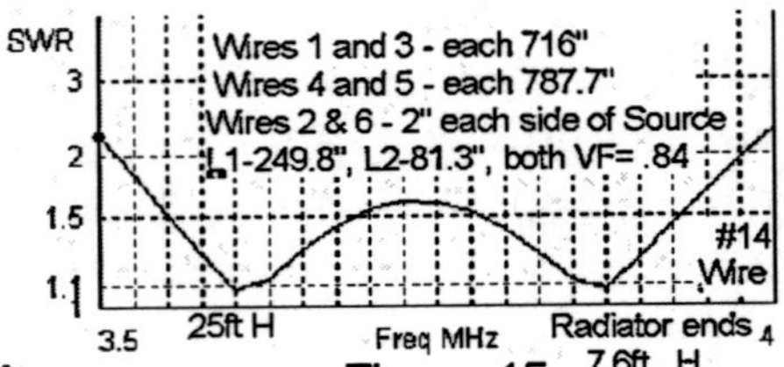

Figure 15

In order to determine what the SWR would be for a single Inverted V mounted at the same height of 8' feet, the SWR plot of Figure 16 shows the resulting SWR of approximately 5/1 at the band edges. The SWR plot of Figure 17 shows substantial improvements to this SWR.

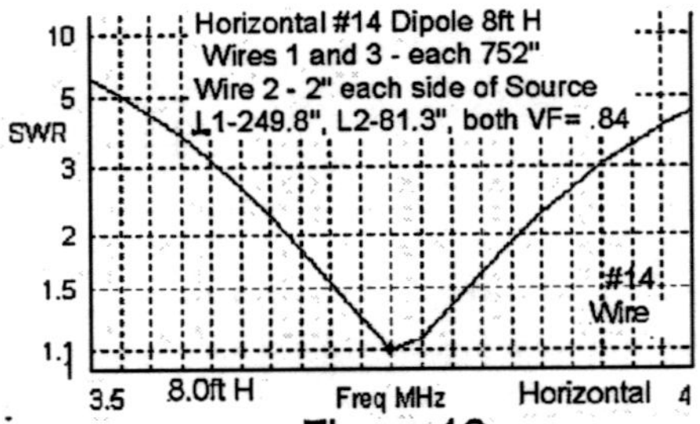

Figure 16

This antenna was lowered and became a horizontal configuration at a height of 8' feet above ground and the resulting SWR plot of Figure 17 shows some improvement in the SWR, even though the antenna is strictly horizontal. Although the SWR at the band edges have not improved, the center bandwidth is now at an SWR of 1.4/1, which is a figure that could not be met by other Broad-Banded arrangements.

The Far-field RF gain and loss chart of Figure 18 shows gains of -2.88, -2.23 and -1.74dBi, across the 3.5 to 4MHz bandwidth. Although these values are quite low due to the antenna height, there are gains with relatively low losses of 0.19, 0.08 and 0.07dB for the "Series-section Matching Lines". Since these lines are 27.6' feet in total length, most of the losses should be considered part of the downward transmission line and not necessarily considered part of the matching lines.

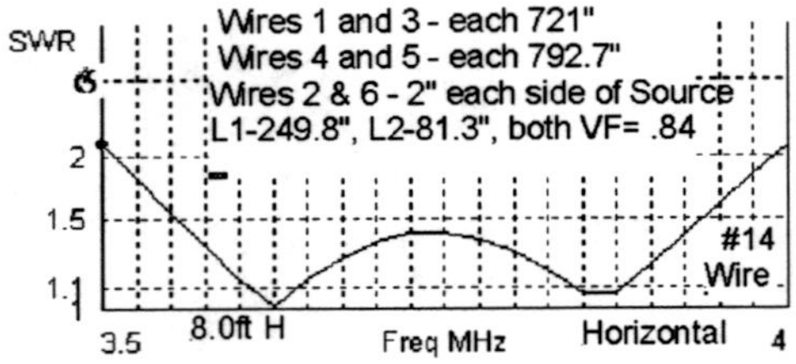

Figure17

Although the RF gains are below 0.0dBi into the minus area, this antenna should perform reasonably well and certainly not require any Antenna-Tuner use and because of the low losses, there cannot be any coaxial cable heating. This is an antenna that should be classified as a "Super Dual Horizontal Antenna".

Far-field Gain & Losses at 89.8°			
8.0ft.H	3.5	3.75	4.0MHz
W/o loss	-2.69	-2.15	-1.67dBi
With loss	-2.88	-2.23	-1.74dBi
Coaxloss	0.19	0.08	0.07B

Figure 18 Horizontal

160m Dual-Inverted V mounted at 90' ft.

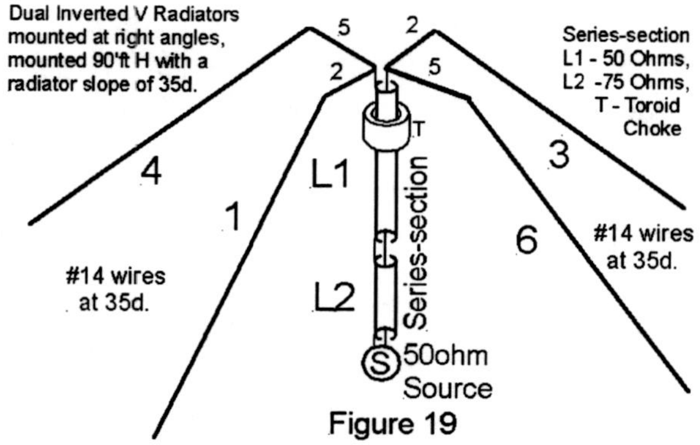

Figure 19

A 75-80m "Dual-Inverted V" antenna was scaled to the 160m band and adjusted in height to 90' feet. It still uses #14 AWG wire and has a slope of 35°, as shown in Figure 19. At a 35° slope, the lowest radiator ends of Wires 4 and 5 will be at a height of approximately 16.7' feet. Remember that this antenna which is mounted at 90' feet in the 160m band is equivalent to having a similar 75-80m antenna mounted at 45' feet.

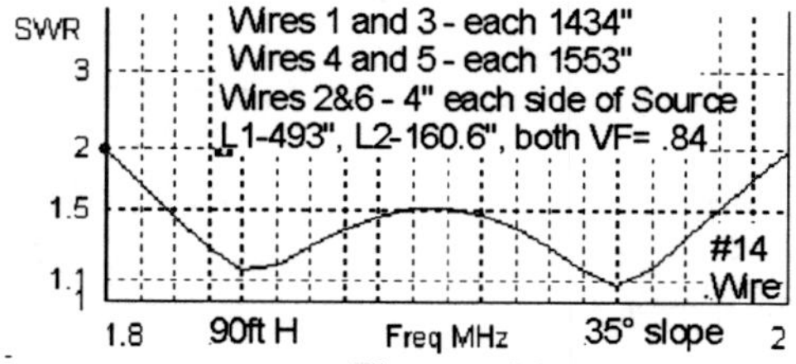

Figure 20

The resulting SWR plot of Figure 20 shows that this antenna has an SWR of 1.5/1 across the entire 160m band from 1.8 to 2MHz and it only deteriorates to an SWR of 2/1 at the very ends of the bandwidth. It is very difficult to obtain good SWR readings at reasonable heights in this band and this has an excellent SWR bandwidth.

The Far-field RF radiation gains, as shown in Figure 21 are very good when the "Series-section Matching Line" losses are eliminated and have values of 4.17, 4.46 and 4.71dBi. These are quite good values but because of the "Series-section" coaxial line length of over 54' feet, these losses distract from these gains down to 3.92, 4.36 and 4.60dBi. These losses and gains could be assigned to be part of the transmission line, while the matching of the "Series-section Line" only constitutes 0.15dB or less.

Far-field Gain & Losses at 89.8°			
90ft.H	1.8	1.9	2.0MHz
W/o loss	4.17	4.46	4.71dBi
With loss	3.92	4.36	4.60dBi
Coaxloss	0.25	0.10	0.11dB

Figure 21 35° slope

If you would care to eliminate the "Series-section Matching Lines", as has been done in Figure 22, you will eliminate whatever RF losses exist but instead, you will have to add transmission line coaxial cable which may have losses approaching those of the "Series-section Matching Lines". Nevertheless, you may not feel that the matching lines are providing sufficient improvement to warrant their use. When the lines were eliminated, Figure 22 shows that the SWR has deteriorated in the band center from a value of 1.5/1, to a value of 1.75/1. If you feel these lines are not necessary, you can eliminate them and place your Source at the juncture of Wire 2. One advantage in doing this is that the Source can be connected directly to a 50/50 Balun and not be a bit concerned about any line lengths in the Balun affecting the antenna performance in any way, except for its internal losses. This is an antenna that performs extremely well in the 160m band, providing excellent SWR values and very good gains. You would be very hard-pressed to find an antenna at 90' feet, with a 35° slope, that can perform anywhere near the performance of this antenna, so once again we have a "Super-Dual Inverted V Antenna" that you will be proud of.

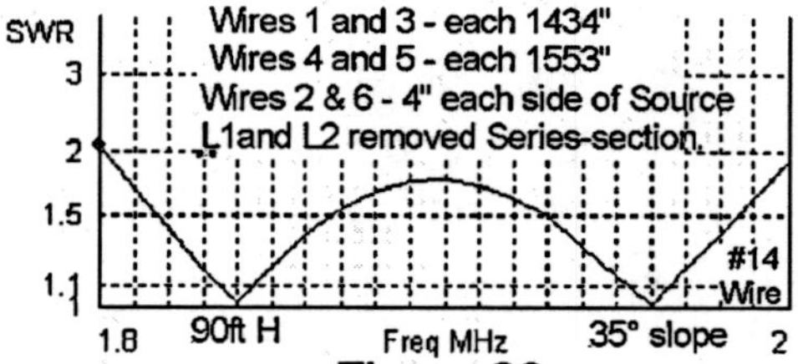

Figure 22

160m Dual-Inverted V mounted at 60' ft.

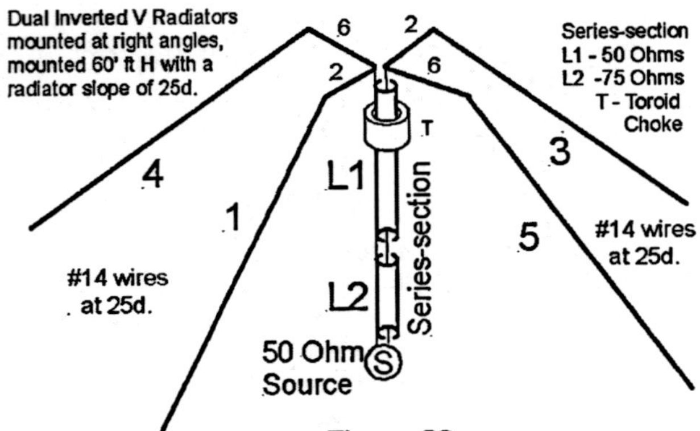

Figure 23

The previous antenna, which is mounted at 90' feet, was reduced in height to 60' feet and in doing this the radiator slope had to be changed from 35° to 25° in order to keep the radiator ends from touching ground. Wires 4 and 5 radiator ends are now 7' feet above ground, using the configuration in Figure 23.

At a height of 60' feet and a slope of 25° an excellent SWR was obtained, as shown in Figure 24. The SWR across the band center is 1.5/1, while the band edges are still at an SWR of 2/1. It has become obvious throughout this chapter that no matter what height or radiator slope is used, the SWR for the "Dual-Inverted V" antennas remain excellent and with very low SWR's. In contrast, none of the "Critically-Coupled" antennas have such characteristics.

Far-field Gain & Losses at 90°			
60ft.H	1.8	1.9	2.0MHz
W/o loss	2.38	2.81	3.20dBi
With loss	2:11	2.68	3.09dBi
Coaxloss	0.27	0.13	0.11dB

Figure 25 25° slope

The Far-field RF radiation chart of Figure 25 shows gain values of 2.11, 2.68 and 3.09dBi across the 160m band, from 1.8 to 2MHz. The coaxial losses for the "Series-section Matching Lines" remain substantially the same.

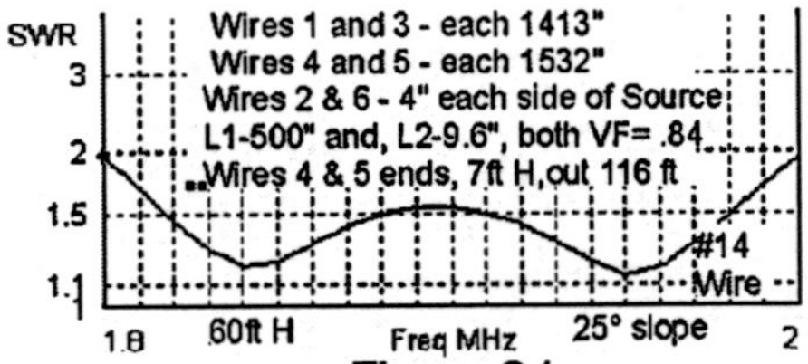

Figure 24

The antenna was lowered from 60' feet to a height of 45' feet and the antenna slope had to be altered to a value of 17° in order to have the radiator ends of Wires 1 and 3 at a height of 10.6' feet, while Wires 4 and 5 are at 8.6' feet, well out of reach.

The SWR of this antenna, as shown in Figure 26, is almost identical to the previous antenna and this seems to be a consistent trend. The SWR remains the same, but the Far-field RF radiation gains of the chart of Figure 27 shows gain values of 1.16, 1.77 and 2.19dBi. At a height of 45' feet, the coaxial losses have remained substantially the same as the previous antenna.

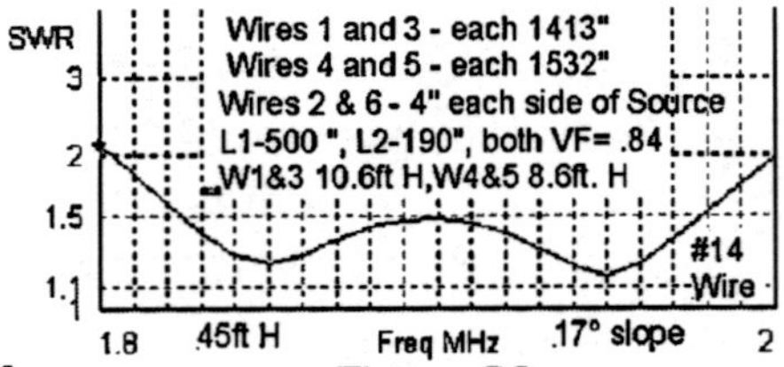

Figure 26

Far-field Gain & Losses at 89.9°			
45ft.H	1.8	1.9	2.0MHz
W/o loss	1.45	1.90	2.31dBi
With loss	1.16	1.77	2.19dBi
Coaxloss	0.29	0.13	0.12dB

Figure 27 17° slope

There should be little doubt that the "Dual-Inverted V" antennas are excellent ways to achieve Broad-Banding, with very low RF losses. No matter what height the antenna is mounted at, it appears to provide excellent SWR values across the entire bandwidth of either the 75-80m or the 160m band's. These antennas are simple to construct and have very low material costs and neither of the radiator wire lengths or the "Series-section Matching Line" lengths are critical. Overall, if you have the area to mount these antennas, they will perform exceedingly well and have some of the lowest SWR values considering the radiator slope.

Super Max- Dipole Antennas

By Edward J. Shortridge W4JOQ

Chapter 6

Super-Double Broad-Banded Antennas

Introduction

Many of the antennas in the previous chapters of this book have introduced a variety of methods of achieving Broad-Banding and although they utilized the "Critical-Coupling Method" of obtaining the best bandwidth, the use of "Dual-Inverted V" radiators has taken the improvement in the SWR to a new higher-level of performance.

The "Dual-Inverted V" achieves its Broad-Banding, by having the 2 radiators mounted at right angles, in effect isolating them from each other and eliminating any "Mutual Inductive Coupling" and although they have a common Source and they do not depend on "Critical-Coupling", they can achieve excellent SWR bandwidths.

Once these radiators are isolated from each other, one is resonated towards the upper end of the desired bandwidth, while the other is resonated to the lower end and thus creating a low SWR. The resulting Source impedance tends to be closer to 60 Ohms, rather than 50 Ohms, requiring "Series-Section" matching.

The improved SWR bandwidth is typically in the vicinity of 1.5/1 across the main portion of the bandwidth and deteriorates to approximately 2/1 at the band edges. This is a great improvement and can be achieved when the antenna is mounted at a variety of heights. When "Critical-Coupling" is used, invariably the SWR bandwidth deteriorates drastically as the antenna is lowered. On the other hand, the "Dual-inverted V" has good SWR bandwidths at all antenna heights.

By having these two radiators resonating at the band edges has put this antenna in a position for further SWR improvement by using an additional resonant circuit, resonating at the band center. This resonant circuit can take different configurations, but in all cases it is intended to resonate at the center of the bandwidth.

In order to obtain this improved SWR, which will be referred to as "Double Broad Banding", the radiators are separated in frequency bias in a slightly greater amount, with this new resonant circuit maintaining its predominance in the band center. In doing this, the SWR at the band edges can be taken down from SWR values of 2/1 to an improved value of 1.5/1. In fact and in almost all cases the SWR is less than a 1.5/1 across the entire bandwidth when operating in the 75-80 or the 160m bands. This is a significant improvement which can be achieved at all mounting heights, using a radiator slope of approximately 35° at the higher mounting heights and using a reduced angle as the antenna is lowered in order to prevent the radiator ends from not only touching the ground, but being raised to a height of 7.5' feet or greater, to have these ends well out of physical reach.

Since the Dual-Inverted V radiators have a common Source, this limits the types of Broad-Banding configurations that can be used. As of this date, the only configurations found that are practical are the "Sx" which uses a short 450 Ohm Ladder/Window-Line, combined with a large resonating Fixed Capacitor. The second "St" configuration utilizes a 50/200 Ohm Balun, which is connected in a reverse fashion internally and so that the 50 Ohm end of the Balun is connected to the radiators, while the 200 Ohm end is connected to a shorted 1/4 wavelength 50 Ohm resonant line which has a 50 Ohm Source tap at about 20% from the shorted end and the total 1/4 wavelength line length is reduced by the Velocity Factor of 0.84. Either of these methods can achieve an SWR of 1.5/1 across the entire 75-80m or the 160m bands and these values will not deteriorate when the antenna is mounted at different heights. These "Super-Double Broad-Banded" antennas have SWR bandwidths far superior to anything generally available, they are easy to construct and have associated reduced cost while some versions are extremely efficient.

160m "Sx" Double Broad-Banded Dual Inverted V antenna

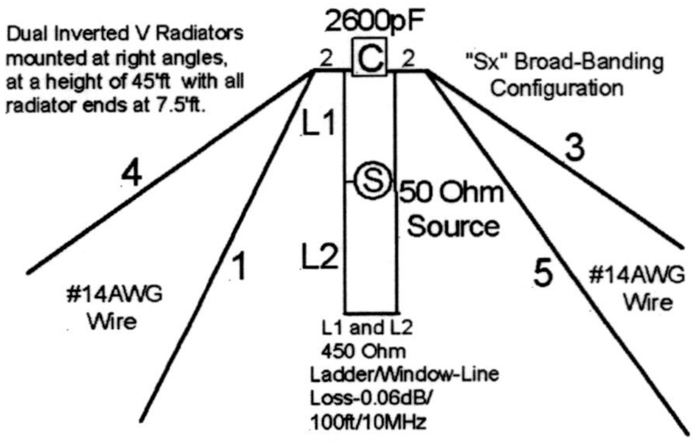

Figure 1

The first antenna to be shown is a Dual-inverted V mounted at 45' feet above Average-soil, with the radiators sloping at approximately 35° and each of the 4 radiator ends extending down to a height of 7.5' feet above ground, as shown in Figure 1. This is the typical Dual Inverted V, as shown in Chapter 5. This configuration lacks the Double Broad-Banding and is the starting point for adding the selected configuration.

The first configuration to be used is the "Sx", which consists of a parallel resonant circuit inserted at the Feed point located at Wire 2. This is the very basic Broad-Banding configuration referred to as the "S", which has been explained in detail in the previous book titled "Super-Dipole Antennas". When this basic configuration is used, the Capacitor in parallel with the inductor must be defined.

The Capacitor can be a single fixed value discrete component or possibly an Open-ended transmission line, which can provide the same capacitive reactants, but in this case a discrete Fixed Capacitor of 2600pF is used. The inductor can be defined as a wound coil with the proper inductance, however in this case it was defined as a Shorted 450 Ohm Ladder/Window-Line, having a total length of 67" inches. This length was divided into two sections, with line L1 being 7" inches, while line L2 is 60" inches. A 50 Ohm Source was placed at the junction of these two lines, in order to achieve the best SWR bandwidth. The impedance across the Capacitor "C" for this antenna happens to be 60 Ohms and it became necessary to tap down on this line to this juncture of L1 and L2, in order to achieve the best SWR bandwidth.

Other Capacitor or inductive values can be used, however it was found that the values selected appeared to provide the best SWR bandwidth, while others seemed somewhat degraded. This is not to say, that this is the only combination that can provide the best SWR, since not all combinations were examined.

The "S", in essence, should present impedance to the radiator feed point such that the reactive components are equal to, but reversed in phase in order to compensate for reactance at this Feed point, increasing the impedance or decreasing it will have effects on the shape of the SWR plot. When this antenna is mounted at a height of 45' feet above Average-soil and all the wires and Line lengths are adjusted for the best SWR bandwidth, an excellent SWR of 1.5/1 or better is obtained as shown in Figure 2. This SWR is far superior to what is generally obtained and could easily become a reference point for SWR achievement.

Instead of specifying a particular angle of radiator slope, such as 35°, this antenna which was mounted at a height of 45' feet, has the Inverted V radiator ends mounted at a height of 7.5' feet, well out of reach. Since each of the 2 radiators have different lengths the resulting angles will be different and the 7.5' foot mounting height for the radiator ends will differ in the distance where they are extended outward, but there mounting heights will be uniform.

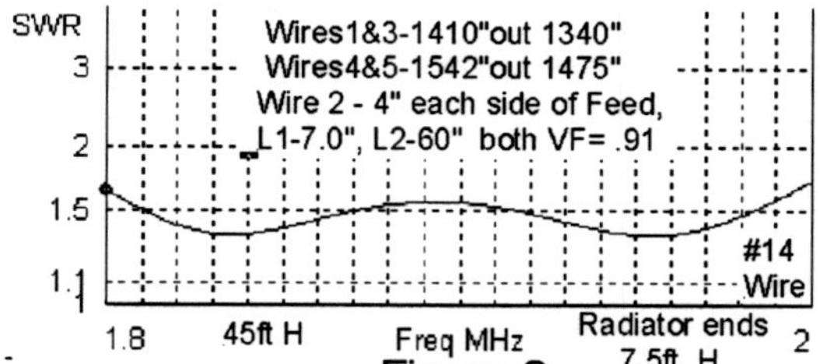

Figure 2

Far-field Gain & Losses at 89.8°			
MHz	1.8	1.9	2.0MHz
W/o loss	1.29	1.62	1.92dBi
With loss	1.26	1.56	1.88dBi
Line loss	0.03	0.06	0.04dB

45ft. H Radiator ends 7.5"H

Figure 3

The Far-field RF radiation gain and loss chart of Figure 3 shows RF gains of 1.26, 1.56 and 1.88dBi, which seems typical for Inverted V antennas in the 160m band mounted at a height of 45' feet. Due to the lack of coaxial cable or long lengths of Ladder/Window-Line, the RF losses have been drastically reduced to 0.03, 0.06 and 0.04dB, which are exceptionally low, making this antenna one of the most efficient "Double Broad-Banded" antennas that has been modeled. You would be hard-pressed to find another antenna with such low SWR bandwidth values and with an efficiency that is rarely achieved by other antennas.

Raising this antenna to a height of 60' feet, the radiator ends are maintained at a height of 7.5' feet above ground and of course the wire dimensions were modified, along with the lengths of lines L1 and L2, in order to maintain a good SWR bandwidth, as shown in Figure 4. This bandwidth still averages at 1.5/1 SWR. It might have been slightly better if the radiator ends were continued to be placed higher and not increased and sloped to a greater angle to a lower height of 7.5' feet above ground.

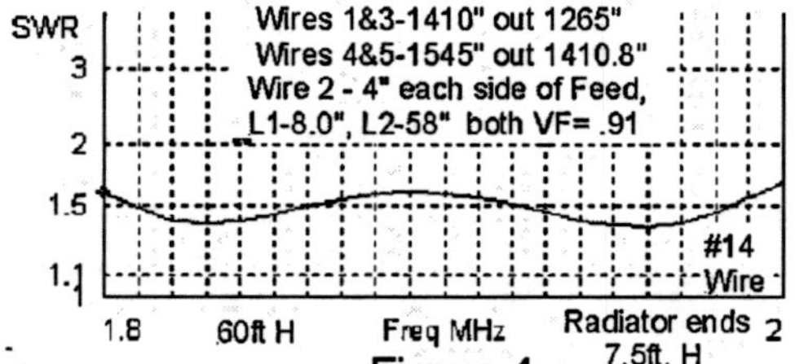

Figure 4

The antenna was further increased in height to 80' feet and the radiator slope was increased in angle to bring the radiator ends to the same height of 7.5' feet above ground. The SWR has changed very little, even though the radiator angle is steeper, as shown in Figure 5.

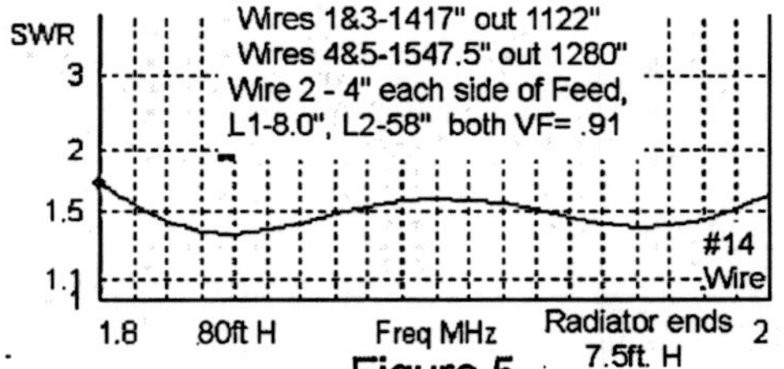

Figure 5

At this higher mounting height, the Far-field RF gains have increased to 3.18, 3.33 and 3.49dBi across the 160m band, as shown in the chart of Figure 6. The RF losses are still maintained at low values of 0.03, 0.06 and 0.04dB across the entire 160m band.

Far-field Gain & Losses at 89.8°			
MHz	1.8	1.9	2.0MHz
W/o loss	3.21	3.39	3.53dBi
With loss	3.18	3.33	3.49dBi
Line loss	0.03	0.06	0.04dB

80ft. H Radiator ends 7.5"H

Figure 6

This 160m "Sx" Double Broad-Banded Dual Inverted V antenna has set a new standard for excellent SWR and the efficiencies that are extremely far above average. This antenna is easy to construct, it has distinct Line Tap capabilities that can be adjusted for the best match and SWR adjustment. It has set a new standard and should easily qualify as a "Super Max-Dipole Antenna" and should be also elevated to a Premier status.

75-80m "Sx" Double Broad-Banded Dual Inverted V antenna

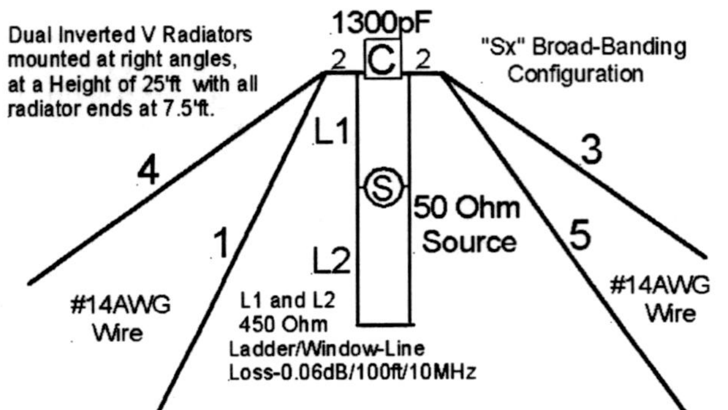

Figure 7

The previous 160m antenna was scaled to the 75-80m band and in doing so the mounting height was reduced to approximately 1/2 or approximately 22' feet above ground. It was readjusted for a height of 25' feet and the radiator ends were once again placed at a height of 7.5' feet, as shown in Figure 7. The scaling process, as used in the modeling program, reduces all of the wire and transmission lines to approximately 1/2 of the former length but the capacitor value of 2600pF had to be changed manually, to a value of 1300pF.

On a percentage basis, the 75-80m band is wider than the 160m band and often it becomes more difficult to achieve an equal SWR bandwidth and that was the case with this antenna. Nevertheless, after considerable manipulation of the wire lengths and the lengths of lines L1 and L2, an almost identical shaped SWR response was achieved as shown in Figure 8. As with the previous 160m antenna, the SWR averages slightly below a value of 1.5/1.

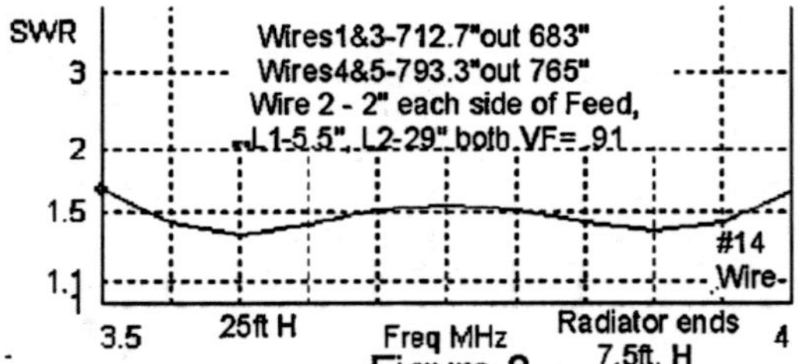

Figure 8

The Far-field gain and loss chart of Figure 9 shows RF gains of 1.66, 2.07 and 2.42dBi, while the RF losses have remained very low at 0.02, 0.05 and 0.02dB across the 3.5 to 4MHz bandwidth. This is still a phenomenally efficient antenna, which is very difficult to duplicate at such a low height of 25' feet.

It is very difficult to find such good performance at low antenna heights; however these lower heights are probably more useful and capable for the average Amateur radio operator to obtain and therefore several low heights are continued to be shown. Therefore, this antenna was increased in height by 10' feet to a height of 35' feet, with the radiator ends lowered to 7.5' feet above ground.

Far-field Gain & Losses at 89.9°			
MHz	3.5	3.75	4.0MHz
W/o loss	1.66	2.07	2.42dBi
With loss	1.64	2.02	2.40dBi
Line loss	0.02	0.05	0.02dB

25ft. H Radiator ends 7.5"H

Figure 9

The SWR bandwidths remained exactly the same as if it were mounted at a height of 25' feet, but to achieve this, some of the wire lengths had to be altered slightly, as shown in Figure 10.

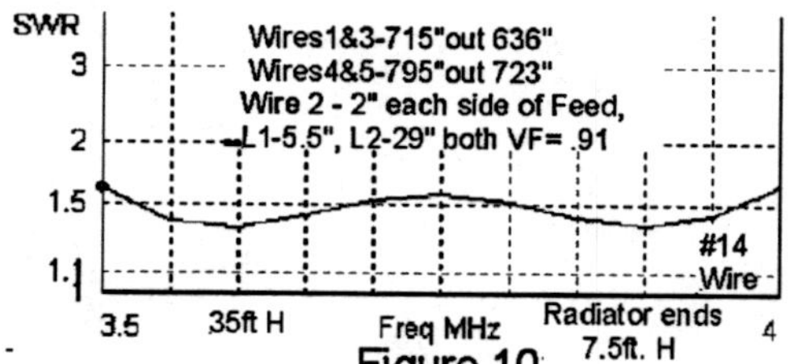

Figure 10

The Far-field RF gain and loss chart of Figure 11 shows gains of 2.76, 3.04 and 3.32dBi, while the losses remained substantially the same. The increase of 10' feet has given slightly greater gains, as would be expected.

Far-field Gain & Losses at 89.8°			
MHz	3.5	3.75	4.0MHz
W/o loss	2.78	3.09	3.35dBi
With loss	2.76	3.04	3.32dBi
Line loss	0.02	0.05	0.03dB

35ft. H Radiator ends 7.5"H

Figure 11

When this antenna was raised an additional 10' feet to a height of 45' feet and the radiator ends are once again lowered to a steeper angle to a height of 7.5' feet, the SWR remained substantially the same, with a slight hint of degradation but this antenna is still averaging close to an SWR of 1.5/1 across the entire bandwidth, as shown in Figure 12.

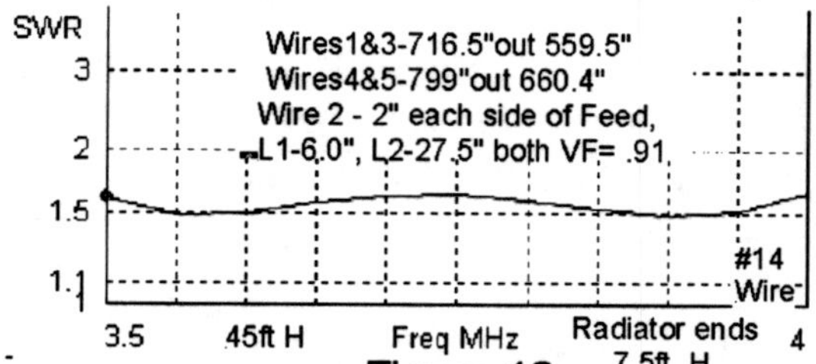

Figure 12

The Far-field RF gains, as shown in Figure 13 have increased further to 3.30, 3.48 and 3.66dBi, while the RF losses remained the same. Besides being an antenna that the gains and losses seemed quite predictable, it is excellent in every respect and this is an antenna that you can be proud to build and use.

Far-field Gain & Losses at 89.8°			
MHz	3.5	3.75	4.0MHz
W/o loss	3.33	3.53	3.68dBi
With loss	3.30	3.48	3.66dBi
Line loss	0.03	0.05	0.02dB

45ft. H Radiator ends 7.5"H

Figure 13

For the final SWR plot, this antenna was raised to a height of 60' feet but in this case, the radiator ends were only sloped down to a height of 20' feet above ground, as shown in Figure 14. This prevented the slope from exceeding approximately 35°. The SWR plot of Figure 14 shows a plot still averaging well below 1.5/1 and slightly better than the previous antenna. There should be little doubt that the "Sx Double Broad-Banded Antenna" can perform with consistent SWR values regardless of their mounting height.

The Far-field RF gain and loss chart of Figure 15 shows increased RF gain values of 4.95, 4.99 and 5.02dBi across the entire 3.5 to 4MHz bandwidth. These are very good gains for this antenna with a height of 60' feet and the losses are substantially the same 0.03, 0.05 and 0.03dB across the bandwidth.

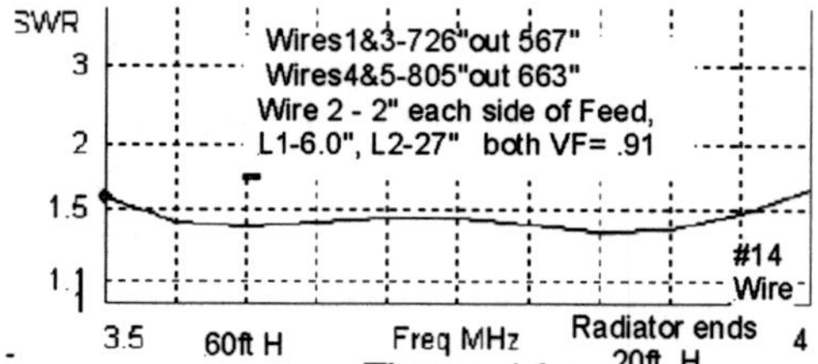

Figure 14

Once again, this shows that this is an outstanding antenna, which has low cost materials, easy to build and especially easy to adjust for an excellent SWR. Since it has separate adjustment points, so that the best SWR can be obtained, it is truly a "Super-Max Sx Double Broad-Banded Antenna".

Far-field Gain & Losses at 89.8°			
MHz	3.5	3.75	4.0MHz
W/o loss	4.98	5.04	5.05dBi
With loss	4.95	4.99	5.02dBi
Line loss	0.03	0.05	0.03dB

60ft. H Radiator ends 20'H

Figure 15

160m "St" Double Broad-Banded Dual Inverted V antenna

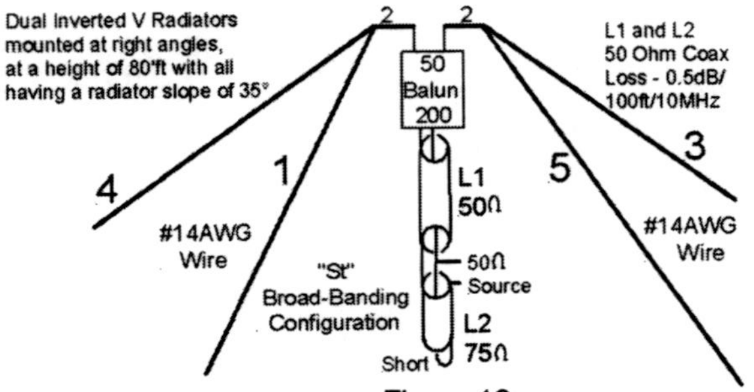

Figure 16

This antenna is a standard Dual Inverted V, adjusted to cover the 160m band from 1.8 to 2MHz and it is mounted at a height of 81.33' feet above Average-soil. The radiator ends are set at a height of 7.5' feet and the "St" Broad-Banded configuration has been added to have it perform as a "Double Broad-Banded" antenna, as shown in Figure 16.

This "St" configuration consists of a standard 50/200 Ohm Balun that has been rotated in its housing so that the 50 Ohm end of the Balun is connected to the Dual radiator Wire 2 Feed-point. The high impedance 200 Ohm end is connected to a 1/4 wavelength 50 Ohm resonant line, that has its lower end shorted.

A tap on this line at the junction of lines L1 and L2 form a 50 Ohm Source. This tap is located approximately 80% down from the upper end and it's 79' foot length acts as the major portion of the downward transmission line, thereby eliminating unneeded length. The lower L2 line, which is 39' feet long, unfortunately cannot be used as a substitute for any portion of the downward transmission line.

In the process of using a 50/200 Ohm Balun, the Feed point impedance of 50 Ohms on Wire 2 is transformed to an impedance of 200 Ohms wherein it is connected to the 50 Ohm line L1. The impedance at the top end of line L1 is transformed to an impedance of 12.5 Ohms, as seen at the Feed point of Wire 2. This in effect takes the 50 Ohm lines L1 and L2 in conjunction with the Balun to have the appearance that it is a 12.5 Ohm line that is being connected to the Wire 2 Feed point.

This has been repeatedly shown in this book and the previous "Super-Dipole Antennas" book that this is a configuration that can cancel out reactive components of the radiators over a wide bandwidth. It generates reactive components that are shifted by a 180° phase, canceling out radiator reactants and in doing this, a much better SWR bandwidth can be achieved.

The standard 50/200 Ohm Balun generally uses 2 windings, each of which possibly uses #14 AWG wire, which will have an impedance of approximately 100 Ohms and the typical length of these windings are in the vicinity of 20" inches. These 100 Ohm lines differ from the 50 Ohm lines used for L1 and L2 but due to their relatively short length compared with the total length of lines L1 and L2, which has a length of approximately 108' feet. These Balun lines tend to have a minimal effect on the 75-80m band and especially on the 160m band.

When this antenna is mounted at a height of 80.33' feet above Average soil, having a 35° radiator slope and adjusted for the best SWR, the plot of Figure 17 is achieved. This SWR averages lower than an SWR of 1.5/1 across the entire 1.8 to 2MHz bandwidth, but it is difficult to obtain a symmetrical shape SWR plot and as a result, the upper frequency end of the bandwidth generally has a lower SWR than that of the lower end. Adjustments have been made to make it symmetrical but these changes generally end with a slightly deteriorated SWR that is symmetrical. Therefore, whenever a Balun is placed on a 1/4 wavelength resonant line, a slide on symmetrical pattern is apt to result.

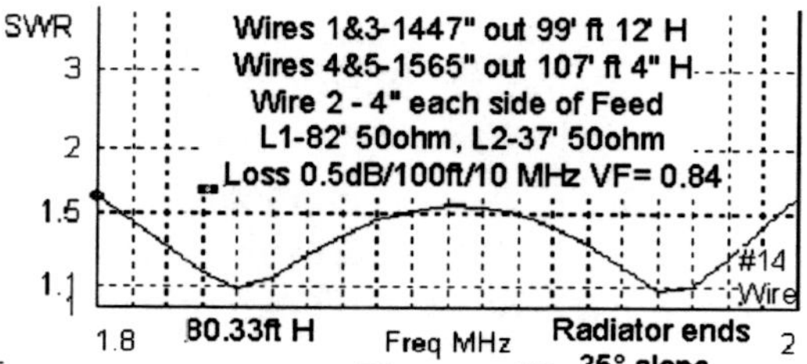

Figure 17

There is little doubt that the SWR, as shown in Figure 17, is every bit as good as that achieved when using the "Sx" Broad-Banding configuration and you might say that the improved SWR at the high end of the bandwidth is actually superior.

The Far-field RF radiation gain and loss chart of Figure 18 shows gains of 2.85, 3.25 and 3.21dBi. These values are relatively low.

The RF losses are shown as 0.45, 0.60 and 0.88dB, which is considerably higher than that of the "Sx" configuration. The "Sx" had no resonant line that could form part of the transmission line, whereas this "St" has a length of 82' feet for line L1 which in effect is used as part of the downward transmission line and to make a better comparison, at least 0.20dB should be subtracted from these losses. Nevertheless, the use of coaxial cable and the resonant lines will have higher losses and every attempt should be made to avoid such cables if possible.

Far-field Gain & Losses at 89.7°			
MHz	1.8	1.9	2.0MHz
W/o loss	3.30	3.85	4.09dBi
With loss	2.85	3.25	3.21dBi
Line loss	0.45	0.60	0.88dB

81.33ft. H Radiator slope 35°

Figure 18

Please note that Line 2 of the SWR plot of Figure 17 indicates that the radiator Wires 4 and 5 slope down to a height of 4" inches above ground, which is not a practical height and may subject people and animals too high RF voltages, therefore all of the radiator ends were changed to a height of 7.5' feet above ground and stretched outward to once again achieve a satisfactory SWR plot, as shown in Figure 19.

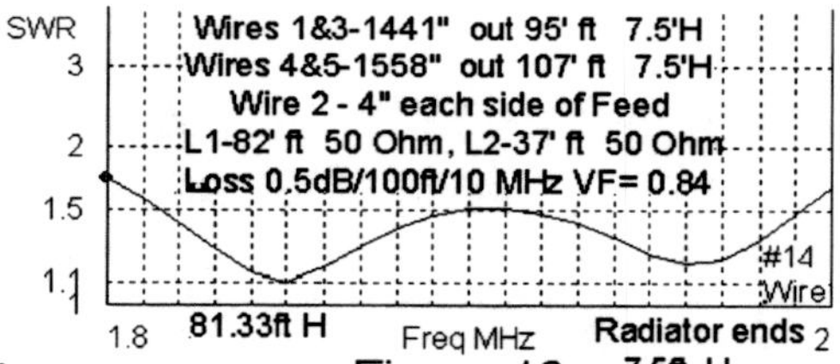

Figure 19

This antenna was raised to a height of 100' feet above ground and the radiator slope has been altered slightly, with the radiator ends at a height of 29' feet. After adjusting the radiator ends the SWR in the 160m band is better than 1.5/1, as shown in Figure 19a.

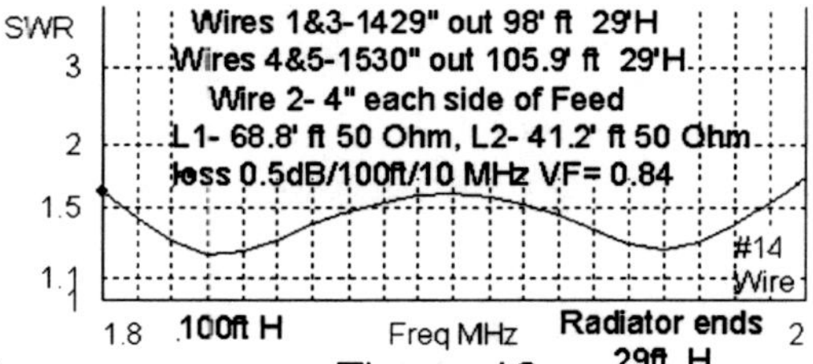

Figure 19a

If this antenna is lowered to 81.33' feet above ground, as shown in Figure 19 and the Source is moved from the resonant line up to Wire 2, then the SWR plot, as shown in Figure 20 can be achieved. This is an excellent SWR and if this Source was to be used, 79' additional feet of transmission line would have to be used to access this higher Source and the 1/4 wavelength resonant line would not have to be tapped but it still continues to be required. Therefore, this higher Source of Wire 2 was not used and Figure 20 is shown only to demonstrate what is possible.

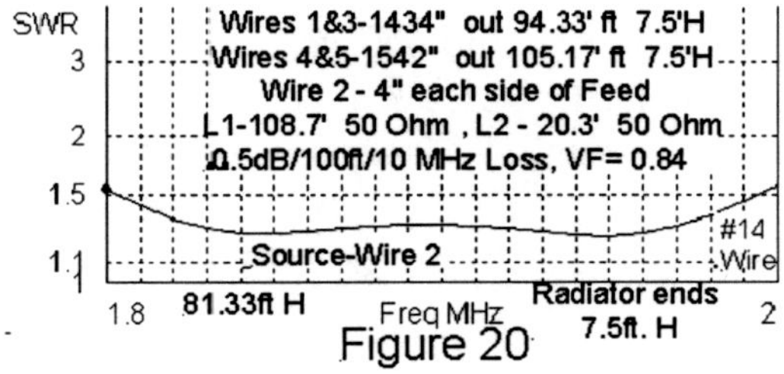

Figure 20

Although this "St" Broad-Banding configuration has higher losses than the "Sx", the losses are not excessive, particularly if you compare them to some of the Broad-Banded antennas currently in use, some of which have losses as high as 1.6dB and the losses of this antenna are far lower and within a reasonable range. If you consider the simplicity of this antenna and it's excellent Far-field RF gains, you could consider it as a good all-around choice, which has all of the ease of adjustments necessary and I would consider it as a "Super-Max Dipole Antenna".

160m "St Unun" Double Broad-Banded Dual Inverted V antenna

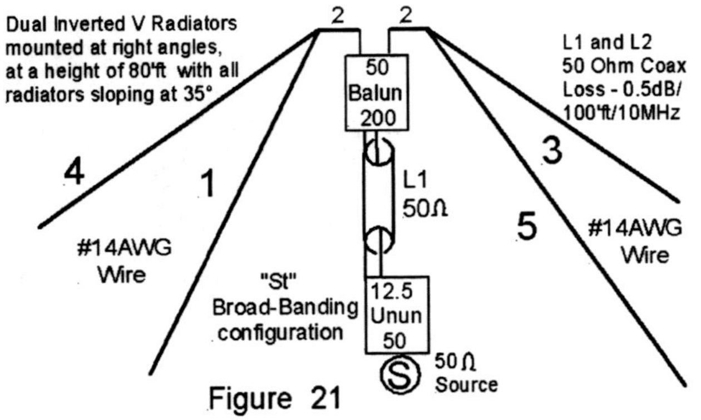

Figure 21

By modifying the "St" configuration with the use of a 50/50 Ohm "Unun", it is possible to use the entire 108.4' foot. length of the 1/4 wavelength resonant Broad-Banding line as a substitute for the entire transmission line. In doing this, only line L1 will be used and the 50 Ohm Source is removed from this location, along with line L2.

If the Source is placed at the lower end of this revised line L1, this Source requires an impedance of 12.5 Ohms in order to achieve satisfactory Broad-Banding. If a 12.5/50 Ohm "Unun" is used it will transform the required Source to a value of 50 Ohms and this arrangement is shown in Figure 21. Such a Balun may not be available but one that is adequate can be quickly implemented from a standard 50/50 Ohm Balun as will be described later.

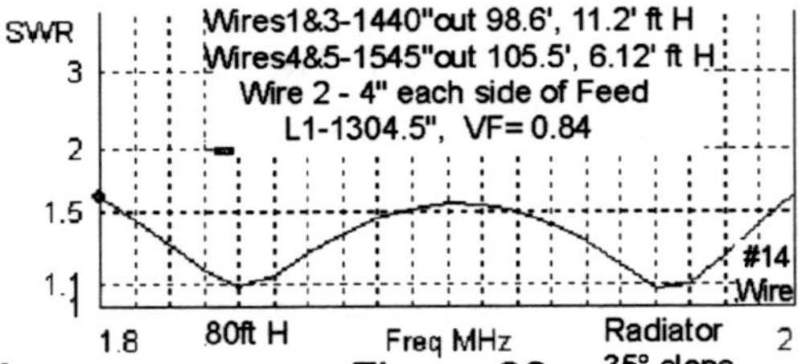

Figure 22

A "St Unun" was used in a Double Broad-Band Dual Inverted V, which is mounted at a height of 80' feet above Average soil and the resulting SWR plot of Figure 22 shows an SWR bandpass which averages well below an SWR of 1.5/1 across the entire 1.8 to 2MHz bandwidth. Although this is an excellent SWR bandwidth, this configuration lacks a means of adjusting the configuration for the best SWR. Nevertheless, this SWR was readily obtained without having to adjust a Tap for the best SWR, as was a necessary requirement in the previous antenna.

Far-field Gain & Losses at 89.8°			
MHz	1.8	1.9	2.0MHz
W/o loss	3.27	3.69	3.97dBi
With loss	2.77	3.02	3.21dBi
Line loss	0.50	0.67	0.76dB

80ft. H   Radiator slope 35°

Figure 23

The Far-field RF radiation chart of Figure 23 shows RF gains of 2.77, 3.02 and 3.21dBi, with RF losses of 0.5, 0.67 and 0.76dB across the entire bandwidth from 1.8 to 2.0MHz. The RF gains seem slightly low and this is due to the RF losses that are higher than desirable. Nevertheless, if you compare this antenna to others such as the "Sx", in which a full length of transmission line must be used and this transmission line will have a loss of approximately 0.2dB, thereby reducing the Far-field RF gain by that amount, you will get RF gains that is more in line with this version of the "St Unun" configuration.

The chart of Figure 24 shows the RF losses that are incurred in this antenna when the 50/200 Ohm Balun, which was connected in a reverse fashion and was modeled using 20" inches of Twin #14 wire for each of the two windings. The Unun was modeled with 20" inches of 50 Ohm coaxial cable for the L1 winding, while the L2 line has only a length of approximately 3" inches or whatever is sufficient to bridge from the input to its output, where it is connected in Series to achieve a 4:1 step up in the impedance to a value of 50 Ohms.

Total Balun & Unun losses		
1.8 MHz	1.9 MHz	2.0 MHz
0.01	0.06	0.01dB
Standard loss in 108'ft of transmissionin line at 160m-0.2dB		

Figure 24

Figure 25 shows the details of connections for the Unun, with line L1 being a standard 50/50 Balun and it can be wound with high temperature small diameter 50 Ohm coaxial cable, or it can be wound with Twin #14 AWG wire, selecting either of these will have little effect on the resulting SWR values finally obtained when using it. Line L2 is a short, approximately 3" inch long coaxial cable that is hooked in Parallel with the input, but in Series with the output, as explained previously. And the standard 50/50 Balun can be used, but it is best if it is constructed with the short coaxial cable of line L2 mounted within a common enclosure.

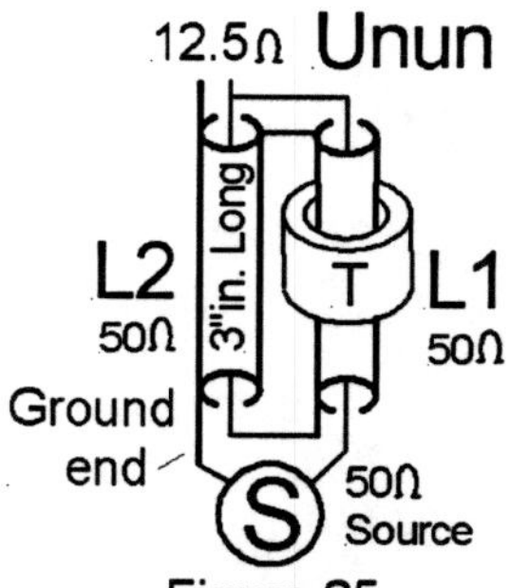

Figure 25

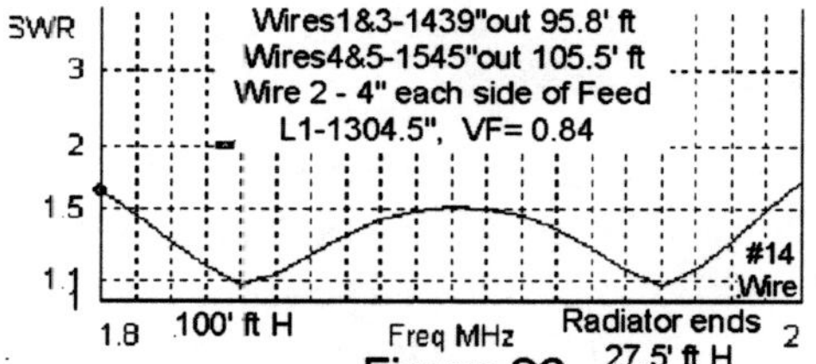

Figure 26

This antenna, which is mounted at 80' feet, was raised 20' feet to a height of 100' feet above ground. In doing this the radiator ends were elevated from 7.5' feet up to a height of 27.5' feet. Due to raising the antenna, the radiator lengths had to be changed slightly in length, but the radiator ends were maintained at 27.5' feet above ground as shown in Figure 26.

You have the choice of selecting this antenna, or the previous antenna that utilizes a 50 Ohm Tap, determine by how your requirements compare with the possible available selections. All of these antennas that have been presented in this chapter have certain advantages and all of them are outstanding antennas that you can be proud of.

75-80m "St" Double Broad-Banded Dual Inverted V antenna

The previous 160m "St" Double Broad-Banded Dual Inverted V antenna as shown in Figure 16 was scaled to the 75-80m band, as shown now in Figure 27. It uses a 50/200 Ohm Balun to increase the impedance from the Wire 2 Feed point, so that the 50 Ohm 1/4 wavelength Shorted resonant line is connected. With this, Broad Banding will occur and the 50 Ohm Source is tapped down on this line to a point that gives the optimum SWR bandwidth across the entire 3.5 to 4MHz band. This antenna is mounted at 60' feet above Average-soil, with its radiator ends sloping down to a height of 22' feet above ground.

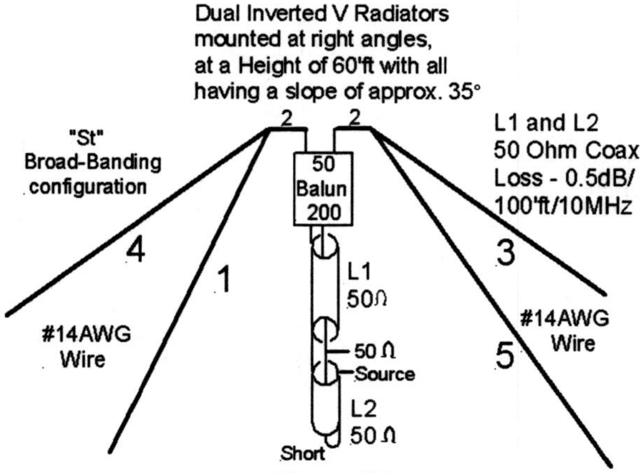

Figure 27

Although the 75-80m band is wider, as figured on a percentage basis than the 160m band, this antenna has achieved an excellent SWR well below an SWR of 1.5/1, as shown in Figure 28. As explained earlier, it appears that whenever a Source tap is placed on a 1/4 wavelength line, the very best SWR achieved across the entire bandwidth tends to be slightly asymmetrical with the best SWR achieved at near the upper end of the bandwidth and no attempt has been made on this antenna to make the bandwidth symmetrical, since in doing it, the overall SWR bandwidth will suffer slightly.

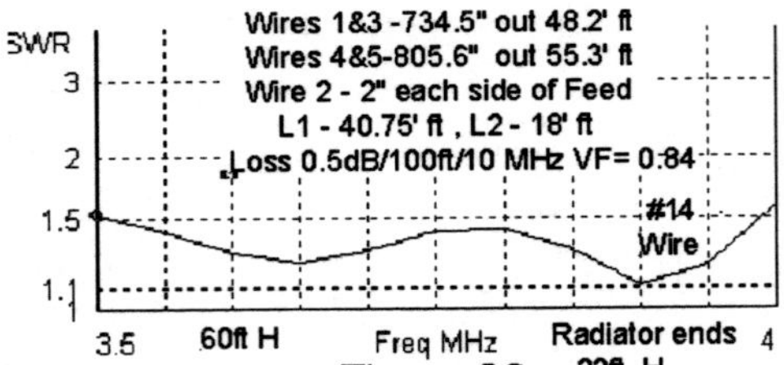

Figure 28

The Far-field RF radiation gain and loss chart of Figure 29 shows that this antenna has extremely good RF gains of 4.62, 4.57 and 4.52dBi, across the 3.5 to 4MHz bandwidth. The RF losses in the coaxial cables are 0.39, 0.52 and 0.57dB across the entire bandwidth and the 40.75' feet length of line L1 acts as part of the downward transmission line and in doing so additional transmission line losses that would normally occur are reduced and you might take these losses and in effect add them to the stated gain figures as additional gain, as shown in Figure 29, in order to make a fair comparison.

Far-field Gain & Losses at 90.0°			
MHz	3.5	3.75	4.0MHz
W/o loss	501	509	5.09dBi
With loss	4.62	4.57	4.52dBi
Line loss	0.39	0.52	0.57dB

60ft. H Radiator ends22' H

Figure 29

When this antenna is lowered 15' feet from a height of 60' feet down to a height of 45' feet and the radiator ends are now set at a height of 7.5' feet above ground, the SWR plot suffers slightly, but still averages less than a 1.5/1 across the majority of the bandwidth, as shown in Figure 30. This is a very minor loss in SWR and is still an excellent SWR for antennas mounted at a height of 45' feet.

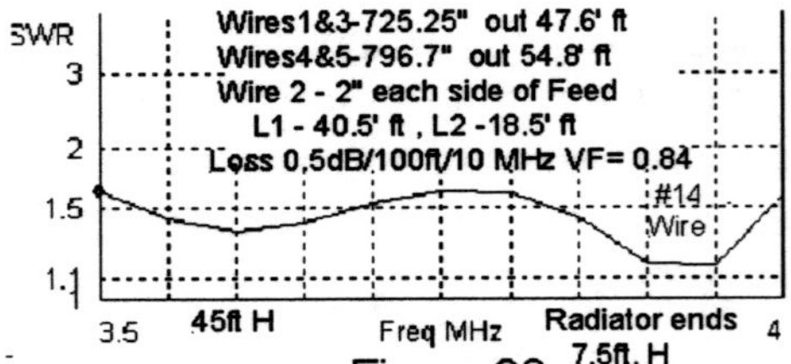

Figure 30

The Far-field RF radiation gain and loss chart of Figure 31 still shows relatively good gains of 2.98, 3.03 and 3.17dBi and the RF losses are slightly higher, as shown in Figure 31.

Far-field Gain & Losses at 90.0°			
MHz	3.5	3.75	4.0MHz
W/o loss	3.35	3.58	3.75dBi
With loss	2.98	3.03	3.17dBi
Line loss	0.37	0.55	0.58dB

45ft. H Radiator ends7.5' H

Figure 31

Lowering the antenna to 35' feet, the SWR plot, as shown in Figure 32 is back to a point where the entire SWR curve is well below an SWR of 1.5/1. Remember, if you were to use the same "St" configuration with a Single Inverted V, the SWR is apt to deteriorate greatly as the radiator height is lowered, but due to the use of Dual Inverted V radiators, this loss at Low Heights does not occur. As of now, these "Dual-Inverted V Antennas" are the only antenna configuration that seems to maintain satisfactory SWR at any mounted height.

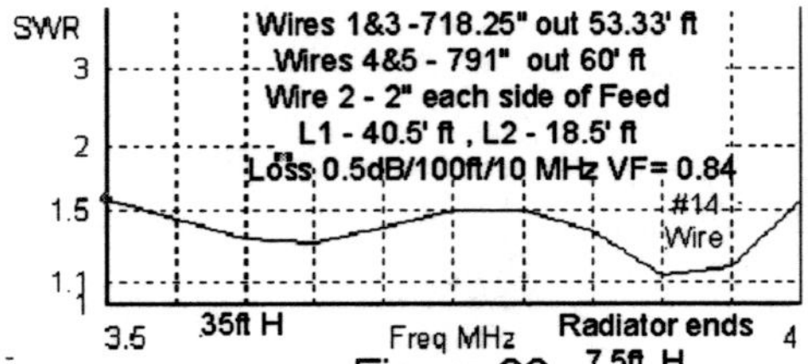

Figure 32

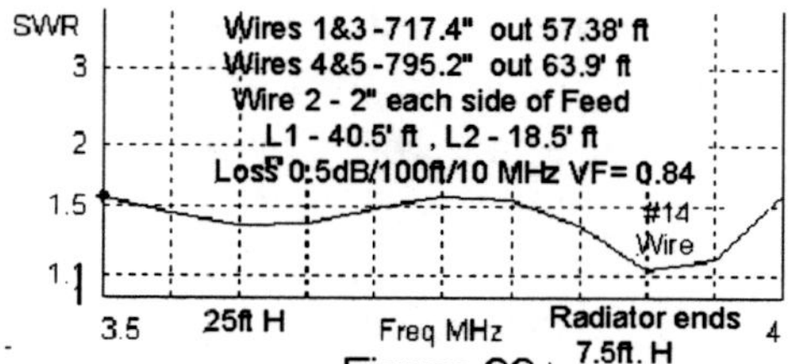

Figure 33

Going one step further, the antenna was lowered 10' feet to a height of 25' feet above ground, with the radiator ends secured at a height of 7.5' feet. The SWR plot of Figure 33 shows substantially the same SWR plot, as that of the previous mounted height.

This antenna, as it has been presented, is an excellent antenna with minimal losses and good gains. It is capable of being adjusted by means of its line L1 and L2 Tap to achieve the very best SWR values. You can be proud to use this antenna and achieve good overall results. By all means, this is a "Super-Max Dipole Antenna".

75-80m "St Unun" Double Broad-Banded Dual Inverted V antenna

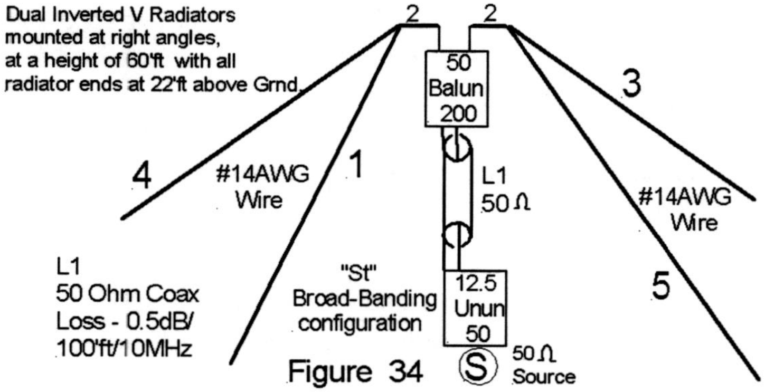

Figure 34

The previous 160m "St" Unun antenna, as shown in Figure 21 was scaled to the 75-80m band and was mounted at a height of 60' feet and the radiator ends were secured to a height of 22' feet above ground, as shown in Figure 34. This antenna utilizes a 50/200 Ohm Balun to transform the 50 Ohm impedance at Wire 2 up to an impedance of 200 Ohms, where it is connected to a 50 Ohm Resonant 1/4 wavelength line. This line gives the very best SWR when the bottom end has a Source impedance of 12.5 Ohms. A 12.5/50 Ohm Unun is utilized to transform this 12.5 Ohm impedance to a 50 Ohm Source.

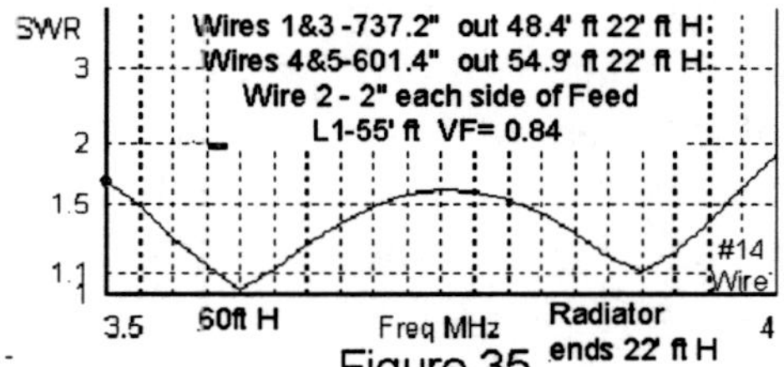

Figure 35

The SWR plot of Figure 35 shows an SWR averaging far less than 1.5/1 across the entire bandwidth but the center of the band has the SWR slightly above 1.5/1. This SWR is more than satisfactory and it has certain portions of the bandwidth dipping down to practically a 1/1 SWR.

Far-field Gain & Losses at 89.7°			
MHz	3.5	3.75	4.0MHz
W/o loss	5.10	5.17	5.18dBi
With loss	4.66	4.66	4.52dBi
Line loss	0.44	0.51	0.66dB

60ft. H Radiator ends 22'H

Figure 36

The Far-field RF radiation gain and loss chart of Figure 36 shows gains of 4.66, 4.66 and 4.52dBi and these are very good gains for this antenna height, while the coaxial losses for line L1 are reasonable at 0.44, 0.51 and 0.66dB across the entire 3.5 to 4MHz bandwidth. Considering that the 55' feet of the line L1 also forms part of the downward transmission line, these listed losses seems very reasonable.

This antenna was lowered to a height of 45' feet but the SWR bandwidth deteriorated at the band center and edges, while the remaining portions are still averaging well below 1.5/1 SWR, as shown in Figure 37. The 75-80m band is considerably wider than that of the 160m band. When a 160m antenna is scaled to the 75-80m band, the SWR bandwidth will suffer slightly. Additionally, there is a difference in the Broad-Banding obtained between having a shorted line L1 with its Source tap on the line, compared to deleting the Tap and instead placing the Source at the bottom end of the resonant line.

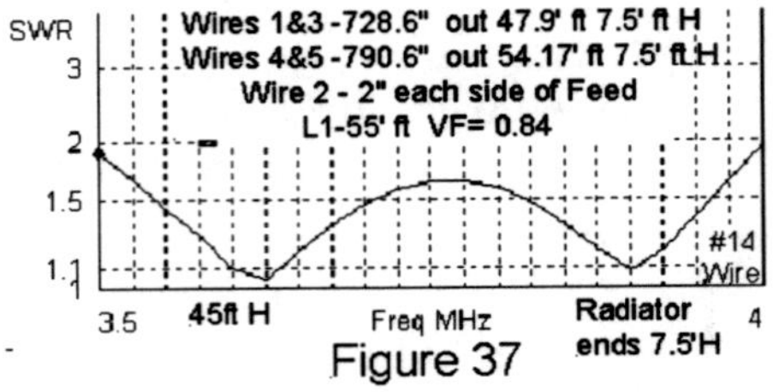

Figure 37

This antenna was lowered an additional 10' feet to a height of 35' feet above ground and the SWR plot Figure 38 shows that the SWR has deteriorated slightly further. Nevertheless, this is still a good SWR but not as low of an SWR as desired.

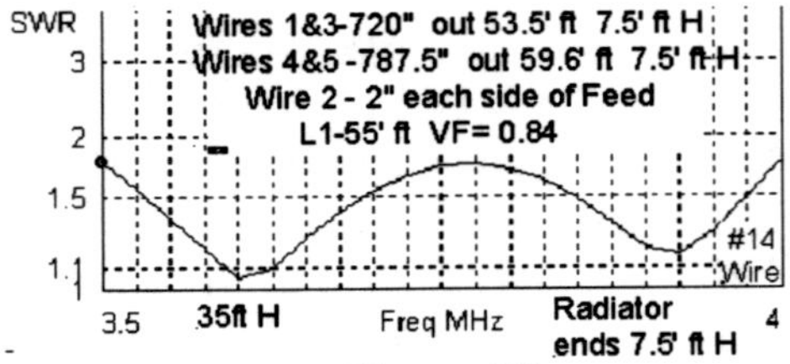

Figure 38

Going one step further, this antenna was lowered to a height of 25' feet and the radiator ends were secured at a height of 7.5' feet above ground, as shown in Figure 39. The SWR has not deteriorated any further and this is a perfectly usable value, especially since it averages less than 1.5/1 across the total bandwidth.

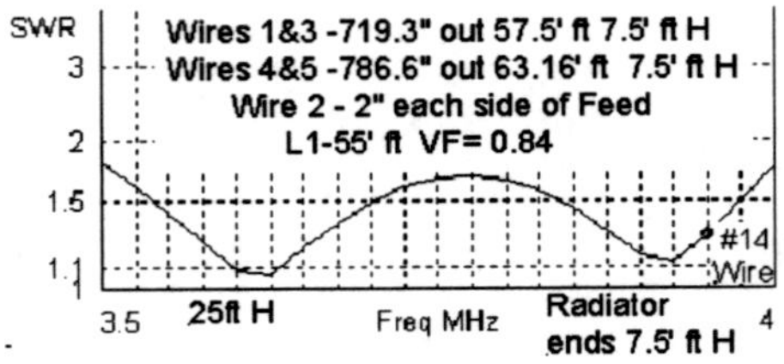

Figure 39

The Far-field RF gain and loss chart of Figure 40 shows gains of 1.40, 1.73 and 2.04dBi across the 3.5 to 4MHz bandwidth. The RF losses have remained substantially the same as those obtained by the antenna which was mounted at 60' feet, as shown in Figure 36.

Far-field Gain & Losses at 89.8°			
MHz	3.5	3.75	4.0MHz
W/o loss	1.84	2.27	2.64dBi
With loss	1.40	1.73	2.04dBi
Line loss	0.44	0.54	0.60dB

25ft. H Radiator ends 7.5'

Figure 40

Although this "St Unun" antenna doesn't exactly match up to the performance of most of the other antennas shown in this chapter, it is still a very good antenna, with an SWR bandwidth far superior to most other antennas and it has RF gains that are very good for an antenna mounted as low as 25' feet above ground. All in all, this antenna has all the qualities to be classified as a "Super Max-Dipole Antenna".

160m "Sz" Double Broad-Banded Dual Inverted V antenna

This "Sz" version of the Double Broad-Banded Dual Inverted V antenna has been a controversial antenna, which has very good SWR bandwidths, but it has some of the higher line losses. It consists of a shorted 1/4 wavelength resonant line which uses 400 Ohm Ladder/Window-Line, which forms a resonant circuit, which provides Broad-Banding.

It consists of two sections, with the shorted line L1 connected to the Feed-point, which is also the Source, as shown in Figure 41. The remaining portion of the 1/4 wavelength open-ended resonant line L2 is also connected to the radiator Wire 2 feed-point. The length of L1 is 8' foot, while that of line L2 is 107.5' feet. This long length of line L2 can hang down and be secured but none of its length can be used as a substitute for any portion of the transmission line.

The impedance at the open end of line L2 is extremely high and there will be a voltage buildup of thousands of Volts, causing the line losses to be excessive even though the standardized losses for this 400 Ohm line is low. Due to these high losses and the fact that it must be mounted somewhere; this antenna was not intended to be in this book. Since the "Sz" configuration is a valid part of the "Unified Broad-Banding Method", it was necessary to include it, even though it lacks credibility as to be classified a "Super Max-Dipole Antenna".

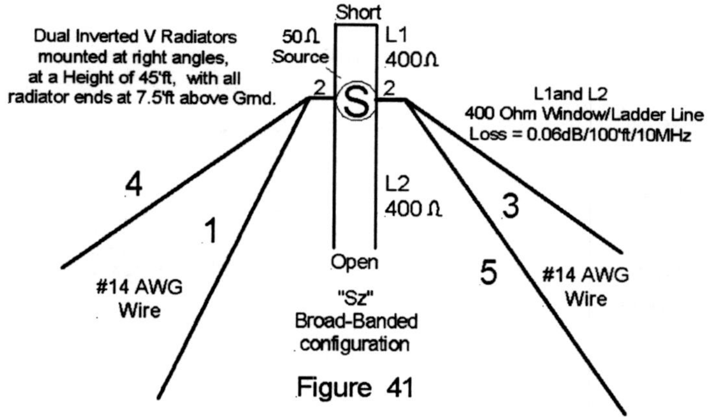

Figure 41

Nevertheless, this antenna when mounted at a height of 45' feet above Average-soil has an excellent SWR, as shown in Figure 42. It seems impossible to achieve asymmetrical bandpass but the one achieved is well below an SWR of 1.5/1, particularly on the upper portion of the bandpass.

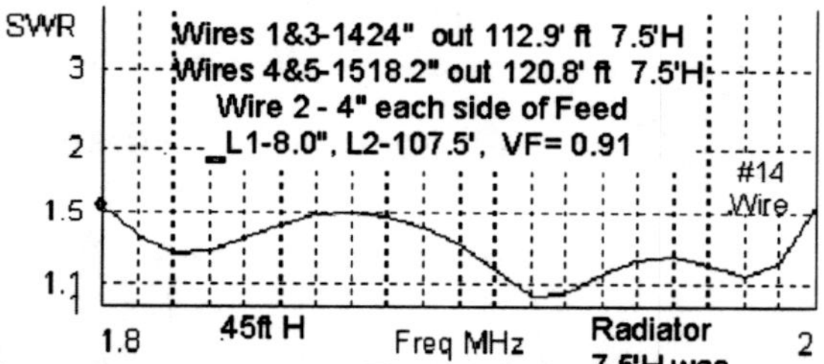

Figure 42

The Far-field RF gain and loss chart of Figure 43 shows gains of 1.17, 1.48 and 1.11dBi across the 1.8 to 2MHz bandwidth. The L1 line losses are shown as 0.07, 0.14 and 0.73dB across the entire bandwidth.

Far-field Gain & Losses at 89.8°			
MHz	1.8	1.9	2.0MHz
W/o loss	1.24	1.62	1.94dBi
With loss	1.17	1.48	1.11dBi
Line loss	0.07	0.14	0.73dB

45ft. H Radiator 7.5'H

Figure 43

The specified 400 Ohm Ladder/Window-Line uses #14 AWG wire and it was selected due to its wire size so that lines L1 and L2 can then be mounted as part of one of the radiators and this would eliminate its mounting problems. It is a difficult task to model such an antenna and questions arise as to what radiator to mount it on and such mounting was therefore abandoned.

Although this antenna may not be used, it's good and bad points have been stressed but its configuration and operating characteristics are of value, giving a sense as to what type of configuration is workable.

Conclusions

This is the end of Chapter 6 and this concludes this book, which covers "Dual Inverted V" antennas and "Double Broad-Banded Dual Inverted V" wire antennas for the 75-80m and 160m bands. Although these are new and unique antennas that can provide amazing SWR bandwidths far beyond the ordinary, most all have low cost of materials and ease of construction. Most have a means of adjusting the configuration for the very best SWR but all of them provide SWR characteristics, that are surprisingly low and wide and well beyond the ordinary. These antennas fulfill a need for antennas that do not require constant antenna retuning each time the frequency of operation is altered.

Glossary of "S" Antenna Configurations

"S" – Parallel Inductor/fixed Capacitor (Basic L/C)

"Sc" – ¼ wL resonant line at Source

"Scr" – (CR) Coupled Radiators

"Sd" – Delta configuration

"Sdd" – Diamond configuration

"Sg" – Gamma configuration

"S-Hz" – High Impedance coupled

"So" – ¼ wL resonant line tapped out on Radiator

"St" – transformer or Balun with 1/4wL resonant line

"St Unun" - "St" with added Unun

"Sx" – Large value Capacitor/parallel Short line

"Sz" – Open-ended 1/4wL with Tapped-down Source

"S-Zepp" – Zepp form of "S-Hz" coupling

wL = wavelength

CPSIA information can be obtained
at www.ICGtesting.com
Printed in the USA
LVOW09s1551190817
545631LV00005B/337/P